THE LAST LIONS

Edited by Don Pinnock and Colin Bell

Foreword by David Quammen

Smithsonian Books
Washington, DC

Serengeti National Park, Tanzania. © Johan van Zyl

Okavango Delta, Botswana. © Hannes Lochner

Etosha National Park, Namibia. © Inki Mandt

Botswana. © Peter and Beverly Pickford, Wildlandphoto.com

Serengeti National Park, Tanzania.
© Johan van Zyl

Serengeti National Park, Tanzania.
© James Lewin

Okavango Delta, Botswana. © Colin Bell

'The generation that destroys the environment is not the generation that pays the price. That is the problem.'

———————

Wangari Maathai

CONTENTS

Foreword		18
Preface		21
01	**Where have all the lions gone?** Dr Andrew Loveridge, Dr Lara Sousa, Dr Samuel Cushman, Dr Żaneta Kaszta & Prof David Macdonald	22
02	**The most charismatic of cats** Dr Natalia Borrego	30
03	**Saving Africa's Lions** Dr Pieter Kat	38
04	**Genetic variability in lions** Dr Laura Bertola	42
05	**The poison problem** André Botha	48

The Sahel

06	**Lions of the north** Prof Hans Bauer	58

Chad

07	**Building a lion stronghold** Chiara Fraticelli	66

Cameroon

08	**Bouba Ndjida National Park** Dr Paul Funston	72

Ethiopia

09	**The elusive lions of Ethiopia** Graeme Lemon	76

Central African Republic

10	**A wildlife refuge comes back to life** African Parks	80

South Sudan

11	**A newly rediscovered migration** Marcus Westberg & African Parks	82

Kenya

The lions of Kenya portfolio		89
Portfolio – Jeffrey Wu		97
12	**Maasai Mara – a fragile Eden** Jonathan & Angela Scott	111
13	**Lions return to Amboseli** Jeremy Goss	122
14	**Lions on the doorstep** Emma Childs	130
15	**Lions, livestock and extinction** Dr Laurence Frank	138

Uganda

16	**Murchison Falls National Park** Dr Paul Funston	144
17	**Tree-climbing lions of Ishasha** Dr Tutilo Mudumba	148

Democratic Republic of the Congo

18	**Lions of hope in the Virungas** Olivier Mukisya, Virunga National Park	156
Portfolio - It's not only leopards that hang out in trees		159

Rwanda

19	**Lions of Akagera** Drew Bantlin, Rob Reid & Andrea Reid	168

Tanzania

The lions of Tanzania portfolio		177
20	**Living with lions in Ngorongoro** Sally Capper	192
21	**Traditional use of lion parts** Jonathan L Kwiyega & Belinda J Mligo	196

22	**Dam and be damned: Stiegler's Gorge** Colin Bell	204
23	**Tanzania: and now for the good news!** Chris & Monique Fallows	210

Angola

24	**Where are Angola's lions?** Dr Paul Funston	215
25	**Lions: quo vadis Angola?** Colin Bell	218

Malawi

26	**The lions of Liwonde** Chris Badger	227
27	**Rewilding Majete** Africa Geographic	234

Zambia

The lions of Zambia portfolio		239
28	**Learning with lions** Thandiwe Mweetwa	246
29	**Life with Lady Liuwa** Andrea & Craig Reid	251

Mozambique

30	**Understanding coexistence** Dr Colleen Begg, Keith Begg & Dr Agostinho Jorge	265
31	**Gorongosa's Lions** James Byrne	278

Zimbabwe

The lions of Zimbabwe portfolio		285
32	**Blood under the carpet** Brent Stapelkamp	296

Namibia

The lions of Namibia portfolio		303
33	**The desert lions of Kunene** Izak Smit & Ingrid Mandt	312

Botswana

The lions of Botswana portfolio		323
34	**The lion miracle in the Makgadikgadi salt pans** Super Sande	348
35	**Eye-cows and the value of lateral thinking** Dr Neil R Jordan	360

South Africa

The lions of South Africa portfolio		369
36	**Bringing back the lions of Lapalala** Peter Anderson	385
37	**The rewilding of Kwandwe** Ryan Hillier	392
38	**Rewilding farmland for lions** Lindy Sutherland	398
39	**Lions return to Babanango Game Reserve** Chris Galliers	407
40	**White lions** Chad Cocking	414
41	**Bones of contention** Dr Don Pinnock	422
42	**Killing the king** Ian Michler	430
Portfolio – Brent Stirton		439
43	**The last lions of Asia** Bhushan Pandya	446
44	**Under-tourism, flight shaming and the scourge of over-tourism** Colin Bell	454
The last word: Rewilding, the planet's last great hope		460
About the authors		470
Acknowledgements		479
References		480
Bibliography		482

© Grant Atkinson

Foreword

The importance of lions includes – but transcends – the cold rational metrics of how a big predator fills its ecological role: consuming prey, cropping herbivore populations, converting meat to energy and growth, etc. Those metrics apply also to such estimable creatures as tigers, leopards, saltwater crocodiles and Komodo dragons, as well as other large flesh-eaters occupying the highest consumer rung within their respective communities.

Lions are different. Special. They are supreme on their landscapes and iconic in human minds. They reside atop what I call the Food Chain of Power and Glory, as registered in the esteem of people all over the planet. They are seen as regal, imperious. It didn't take a movie called *The Lion King*, grossing billions of dollars worldwide, to prove that.

And yet, almost paradoxically, lions also embody the ineffable quality of 'wildness' to a maximum degree. They seem unpredictable and highly inconvenient if you share a neighbourhood with them, and downright scary – they are unquestionably dangerous.

You can read about lions as agents of ferocious menace in the Biblical book of Daniel, chapter six. You can see them portrayed, sublime and beautiful, on the walls of Chauvet Cave, as those walls were painted by some lion-besotted artist roughly 30,000 years ago. You can view them in the brilliant, sometimes searing, photographs of Brent Stirton and others, as sampled in this book. And you can judge for yourself.

So, lions are majestic and formidable, yes. But contrary to appearances, they are also quite vulnerable, both individually and collectively.

Individually, the life of any lion in the wild – cub, juvenile, adult male or female – is a precarious journey amid mortal hazards. Foremost among these hazards for a lion population under natural circumstances are other lions. As the veteran lion biologist Craig Packer once told me: 'The number one cause of death for lions, in an undisturbed environment, is other lions'.

Cubs are often killed by new males taking over breeding privileges with a pride of females. Males kill other males in disputes over those prides. Even females sometimes kill cubs or other females of a competing pride. It's important to note immediately, though, that 'an undisturbed environment' for lions, free of human-caused pressures and threats, is a rare idyll nowadays.

Collectively, lions are vulnerable to long-term population decline caused by loss of habitat, fragmentation of remaining habitat, disappearance of native prey, conflict with farmers and pastoralists, hunting in all forms by humans, and local use and international traffic of their body parts. These are some of the factors that have catastrophically reduced their total numbers and left many small, isolated populations in peril of inbreeding depression and other secondary problems that can push those small populations to oblivion.

Their distributional range once spanned not just most of Africa (encircling the Sahara) but up into Europe, including the Ardèche valley of southeastern France (where Chauvet Cave is located), across the Middle East (as testified by the prophet Daniel, among other sources), and further eastward as far as Balochistan (in what is now Pakistan) and northern India to the very outskirts of Old Delhi.

Those far-flung lions and their offspring are now gone, reduced to a remnant of about 20,000 animals (or maybe as few as 10,000, according to Pieter Kat) existing within a handful of large habitat strongholds and a few score of smaller enclaves scattered across Africa, plus roughly 700 of the Asiatic population holding out in the Gir Forest region of western India. As human population growth has exploded upward, in all those places, lion population size has plunged. It doesn't take a PhD in ecology to comprehend that relationship.

And it doesn't take me, chattering here, to elaborate on the factors that drive the downward trend, because those factors are all addressed, with authority and in detail, by the scientists, conservationists and other lion experts whose voices are gathered in *The Last Lions*.

These deeply informed reports will give you a broad understanding, with some fine local focus, of what has happened and is happening to this great kingly and queenly beast. Equally important, they will alert you to what should happen – what must happen – if our world is to be graced by a continuing presence of wild lions into the indefinite future. You'll read, not just of loss and decline, but of ideas, solutions, projects and efforts to stanch the losses and curtail the decline. You'll read of new ways that ingenious, dedicated people are finding to say Yes to wild lions and No to the prospect of their final extinction, which is a grim possibility but by no means inevitable.

This book will also help remind you that, without wild lions, planet Earth would be a lonelier, uglier, more boring place, a sorely diminished spheroid of rock, carrying humanity into a lionless future. The first step towards averting that outcome is to embrace lions a little more knowingly, in their wondrous and difficult complexity, their majestic inconvenientness, their beauty and their glorious roar. They seem obvious, in a way: Lions, yeah, of course, lions. But we should never take them for granted.

David Quammen

Lions on the salt pan in Etosha National Park, Namibia. © Inki Mandt

Serengeti National Park, Tanzania.
© James Lewin

Preface

The intense, unblinking stare and the flick of a tail of a hungry wild lion leave you in no doubt that, without the protection of a vehicle or firearm, you would be dead meat. For thousands of years these powerful predators at the top of the food chain struck terror in the hearts of our ancestors – and would have eaten a good many of them.

Today, the tables have turned, and we occupy top position. In getting there, we have wrought terrible damage to competitors and prey alike. Lions once dominated most of the Old World but have been pushed to the margins. In Africa they have become locally extinct in 24 countries. In 1970, there were an estimated 92,000 lions in the wild in Africa; the current estimate is around 24,000, most in protected reserves. These lions are highly fragmented across the continent and at risk of extinction in our grandchildren's lifetime.

Motivated by our shared knowledge of Africa and our love of wild creatures, in 2019 we gathered together the writings and photographs of a wide range of people with special understanding of elephants to produce a beautiful book, *The Last Elephants*, which was published worldwide in English and Chinese.

In doing so, we learnt (often the hard way) what it takes to co-ordinate and edit a large-format, 488-page book of that scale. This realisation seemed too good to waste. Linked to growing concern about the state of lions in Africa, we were easily persuaded by our publishers to undertake a similar book for *Panthera leo*. Putting out the word among 'lion' people across the continent, we were humbled and buoyed by their willingness and enthusiasm to come on board to write and help produce *The Last Lions*.

What you have in your hands is a book containing many lifetimes of knowledge, from diverse disciplines, written in an accessible way and supported by some of the world's finest wildlife photographers.

There are chapters on the state of lions in various regions – among others, the Sahel, Ethiopia, Kenya, Tanzania, Uganda, Malawi, Mozambique, Angola, Zambia, Zimbabwe, Namibia, Botswana and South Africa – as well as fascinating explorations of lion behaviour in particular places, such as a passion for climbing trees, a taste for livestock, or dealing with human neighbours. Among the issues explored are hunting, poisoning, the use of lion parts, captive breeding, loss of habitat, rewilding and genetic variability. There's also a glance at the rare lions of the Gir Forest in India.

The book incorporates two magnificent photographic portfolios: one by Getty Images photographer Brent Stirton, well known for his work with the *New York Times, Washington Post* and *National Geographic*; and Canadian Jeffrey Wu, contracted to Nikon and considered one of the most influential wildlife photographers in China. 'Life is not how many breaths you take,' he says of his work, 'but about the moments that take your breath away.'

The foreword is written by American science, nature and travel writer David Quammen. His ground-breaking books include *The Song of the Dodo: Island Biogeography in an Age of Extinctions, Wild Thoughts from Wild Places, Spillover: Animal Infections and the Next Human Pandemic* (written before COVID-19), and *The Tangled Tree: A Radical New History of Life*. His articles have appeared in *Outside* magazine, *National Geographic, Harper's, Rolling Stone, The New York Times Book Review, The New Yorker* and more.

Africa is a massive continent, and in this book we could never cover every one of the lion projects and issues that grace (or plague) the continent, nor include all the good work that NGOs, the private sector and even some government agencies do every day. We had to select what we felt was a fair representation of the good and the bad to give a balanced perspective of the health of lions around Africa. For those who are out there doing amazing work who we have not included, we apologise.

This book is written in the hope that we can avoid the implications of its title, *The Last Lions.*

Dr Don Pinnock, Colin Bell

01

Where have all the lions gone?

Discovering how many lions there are in Africa, where they live and how they're connected is crucial for their conservation and survival.

Dr Andrew Loveridge, Dr Lara Sousa, Dr Samuel Cushman, Dr Żaneta Kaszta & Prof David Macdonald

Large carnivores have experienced rapid population decline and geographic range reductions globally, caused mainly by loss and fragmentation of habitat, conflicts with livestock owners and hunting.[1] **As human populations grow, the conversion of wild habitat to agricultural land is expected to increase fragmentation and species extinction risk across most wildland vertebrates.**

African lions are key to understanding this process. They were once one of the most widely distributed African savanna mammal species, with their geographic range extending from the Mediterranean to the southern tip of the African continent.[2] Range collapse driven by habitat loss and persecution by humans has resulted in the extinction of the northern and southernmost populations and severe fragmentation and population decline across the remainder of the species range.

These threats are predicted to accelerate and become more severe in the next 50–100 years.[3] Contemporary views of species conservation suffer from shifting baseline syndrome, in that modern attempts to halt biodiversity decline often focus on existing geographical ranges and ignore or underestimate historical declines or trends.[4] Failure to appreciate the magnitude and drivers of past loss may lead to an underestimation of contemporary extinction risk. Equally, pessimistic views of past population decline can hamper the realistic assessment of species conservation status and downplay conservation successes and population recovery.

Lions are known to have experienced a geographic range retraction of around 85% since 1500 CE and are currently accepted as numbering around 23,000 individuals.[5] This disconcerting trend has resulted in a reassessment of the International Union for Conservation of Nature (IUCN) Red List status of the West African lion subpopulation as Critically Endangered and focused conservationists on the currently secure – but likely declining – populations in eastern and southern Africa.[6]

In addition to population declines, recent comparisons between contemporary and historical genetic diversity have shown declines in diversity likely owing to fragmentation and loss of habitat connectivity.[7] Together, declining populations and collapsing geographic range – with accompanying

Seenka and Kini, two of a coalition of five from the Ol Choro Oiroua Conservancy, Mara Region, Kenya. © Øyvind Løkka

loss of genetic diversity – threaten the viability of the African lion unless ambitious conservation measures are implemented.

Population size

Accurate estimates of historical African lion population size are absent. A baseline figure of 200,000 is often cited, relating to either the 1970s[8] or earlier in the 20th century.[9] This estimate does not appear to be based on any empirical data or analysis and is likely to be derived from an article by Myers (1975), which speculates that populations could be 'as low as 200,000 or less', having been 'halved since the 1950s'.

In order to generate baseline estimates of present landscape connectivity and size of lion populations, we used published historical sources along with species distribution models approximating the situation in 1970. We chose this decade because (a) there are credible, suitably detailed published sources of information for this period; (b) changes over the 50-year period between 1970 and the present are likely to be informative as human population and development have accelerated over this time, with the sub-Saharan human population doubling between 1975 and 2001; and (c) detailed continent-wide spatial data sets required to generate species distribution models and analyse patterns of landscape connectivity were available for this period.[10]

In 1970, we estimate a population density of 92,054 lions. Our historical population model reveals there may have been ~1,600 lions in sub-Equatorial West Africa in Gabon and southwest Congo Basin, within a sizeable area of connected savanna habitat. This estimate is difficult to verify as there is sparse historical data on population size available for the region. Nevertheless, lions were described as 'widespread' in the Congolese savanna in the 1940s[11] and there are sporadic contemporary records of lions in the Batéké Plateau region of Gabon and northern Congo,[12] indicating that the remnant of a once-widespread regional population may persist. It is suggested that lions from this region are more closely related to southern African populations than those from West and Central Africa,[13] indicating possible southern historical connectivity.

These numbers imply that, compared with current estimates, lion populations have declined by around 70,000 (~75%) in 50 years, or by roughly 1,400 lions a year over five decades. Regionally, they have declined by 93% in West and Central, 65% in eastern, 73% in southern Africa and 95% in the Congo Basin. Despite differing in the analytical approach and temporal scales of analysis, this finding is consistent with declines of 42% over two decades from the mid-1990s estimated by Bauer, H et al. (2015). The estimate of 200,000, when used as a population baseline for the 1970s, is therefore likely a substantial overestimate, though it may be plausible as a baseline for an earlier period.

Regionally, the West and Central African subpopulation has declined by 87% (from 1,600 to ~200), the eastern African subpopulation by 65% (~31,000 to ~11,000) and the southern African subpopulation by 73% (~36,000 to ~9,800). The Congo Basin population, estimated at ~1,600 lions around 1970, has been almost entirely extirpated, declining by 93% to 211 lions currently.

Population connectivity

Analysis of recent historical and current lion landscapes identifies three distinct classes of habitat connectivity. First, core connected habitats are areas that are highly likely to be traversed by dispersing lions and protect breeding populations. Core habitats for lions are almost invariably centred on one or more large protected areas. Second, non-core connected habitats are less permeable to lions, but nevertheless represent critical habitat linkages that could support lions and facilitate movement and gene flow. Non-core habitat usually surrounds areas of core habitat or is formed by smaller protected areas. Finally, we identified corridors that traverse unconnected habitats and provide links that dispersing lions might infrequently use to move between patches of suitable habitat and represent critical linkages for gene flow.

The 1970s lion landscapes, while having contracted by ~60% from the maximum-historical extent, nevertheless consisted of expansive core habitats surrounded by a relatively continuous matrix of non-core habitats. Apart from isolated patches in the far West and Congo Basin regions,

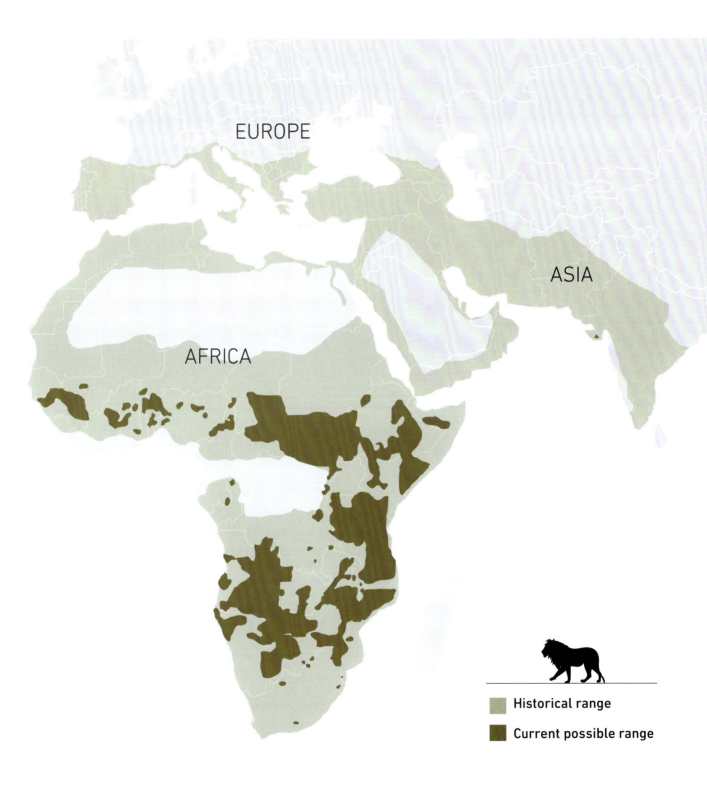

Just 70 years ago there were around 450,000 wild lions roaming around Africa. Today, there only around 23,000 wild lions – a decrease of 95% in just one lifetime.

Where have all the lions gone? ~ **25**

lion range was largely contiguous and connectivity models suggest there was potential for a high degree of dispersal movement across the landscape. By contrast, contemporary lion range is presently significantly more fragmented than 50 years ago, with the overall range declining by ~87% from the maximum-historical extent and ~66% from recent historical range, demonstrating that habitat loss has accelerated markedly.

Of the 34 sub-Saharan countries containing connected lion habitats in 1970, only 26 retain connected habitat currently and, of these, only 25 retain core lion habitat. At a national level, the extent of core lion habitat has almost universally declined and become increasingly fragmented. Many formerly contiguous core areas have been fragmented into small, isolated and less permeable habitat patches.

The West, Central and Congo Basin regions have been most heavily affected, losing from 83% to 90% of their range relative to the 1970s. These regions also showed the highest levels of loss and fragmentation in comparison with maximum-historical range, suggesting that processes contributing to range decline were already taking place by the middle of the 20th century. Eastern and southern regions have suffered lower loss of connectivity overall, losing ~50% of connected habitat, but maintaining several core areas of continuous habitat centred on expansive protected area complexes.

Genetic viability

Species with highly fragmented ranges face a high risk of extinction, with small, isolated populations more vulnerable to environmental and demographic stochasticity.[14] They are more likely to suffer from loss of genetic diversity and inbreeding depression as a result of reduced effective breeding population size and reduced gene flow, with strong correlations between loss of heterozygosity and declines in population fitness.[15] The effects of inbreeding and loss of genetic variability following population fragmentation or isolation have been demonstrated for several large felid populations.[16] This has manifested as deleterious genetic abnormalities affecting breeding success, such as reduced testosterone, sperm abnormalities and cryptorchidism,[17] as well as increased vulnerability to disease as a result of eroded genetic variability.[18]

Recent analysis has shown marked declines in allelic richness, heterozygosity and private alleles in modern, compared with historic, lion populations.[19] Furthermore, modern continental lion populations are more highly structured genetically than in the past, when lion populations were likely regionally panmictic with high levels of male-mediated gene flow, implying that barriers to gene flow exist in the modern landscape that were absent historically.[20] Genetic isolation appears to align closely with observed patterns of habitat fragmentation and conforms closely to the declines in habitat connectivity from recent historic to current times.[21]

Conservation implications

In our models, increasing human population density, more widespread infrastructure and larger urban areas greatly increase landscape resistance in the current landscape compared with that of the 1970s, reducing the permeability of the landscape and the movements of lions. Human population growth in Africa has accelerated since the 1950s, with populations increasing by 59% and per capita resource use having increased by 73% by the 1970s.[22] The population of sub-Saharan Africa doubled between 1975 and 2001 and will double again by 2034.[23] This will likely result in significant clearance of natural habitat for agricultural production, increasing resource extraction and consumption,[24] and increasing urban settlement and transportation infrastructure.[25] This rapid growth of human enterprise has already resulted in severe environmental degradation across much of Africa and will continue to do so.[26] It is expected to accelerate existing trends of habitat fragmentation and isolation of protected areas.[27]

In this process, core connected habitat for lions is being continuously eroded and has been since historical times, with some 26% of African countries losing all core habitat since 1970. Increased human impact is greatly reducing the ability of lions to move through the landscape: areas with currently weak or absent connectivity coincide closely with regions where a high risk of human conflict with lions and elephants (*Loxodonta africana*) is predicted.[28]

Much of the remaining core habitat for lions is centred on sizeable protected areas and the future of lions and Africa's large mammal diversity depends critically on the management of these areas, which are worryingly under-resourced. Many are failing.[29] As such, securing the protection of African conservation areas through greater investment in protected area management is crucial.

Nevertheless, even if core areas are secured, genetic diversity in populations of wide-ranging species such as lions is likely to continue to decline without adequate connectivity between populations. Preserving what remains of lion genetic diversity will require either intensive and resource-expensive meta-population management with animals translocated between isolated core areas to maintain gene flow,[30] or protection of existing movement corridors that allow movement of both lions and other species between patches of natural habitat.

The former approach is widely practised for the management of predator genetic diversity and population dynamics in small, fenced populations in South Africa,[31] but also might be appropriate in regions such as West and Central Africa where lion populations are already irretrievably isolated.

The latter approach – where existing dispersal corridors can be identified, prioritised and protected – is critical for maintaining existing levels of habitat connectivity for lions, gene flow and future genetic diversity. This may yet be a viable strategy in expansive connected landscapes in East and southern Africa and, significantly, would provide genetic and demographic exchange for a wide range of other non-target species.

However, even in regions where landscape connectivity remains, these connections are likely to be highly vulnerable in the face of growing human populations and agricultural activity, particularly in the vicinity of protected areas where human population growth has been shown to be high.[32] Of the two types of connectivity linkage, corridors across unconnected habitats are the most vulnerable to loss, as they generally traverse unprotected, human-occupied lands where lions are less likely to be tolerated and more likely to be killed.[33]

Encouragingly, though, movement data collected from lions fitted with GPS telemetry in the Kavango-Zambezi (KAZA) Transfrontier Conservation Area provides evidence that, at least in this region, these empirically modelled linkages are realistic and still used by dispersing lions.[34] This suggests it is not too late for conservation managers to implement initiatives to secure these corridors through integrated land-use planning exercises,[35] implementation of human-wildlife conflict mitigation strategies,[36] and enhancement of sustainable, wildlife-based livelihoods as part of landscape-scale conservation programmes.[37] Future work should focus on scenario analyses, which can optimise the trade-offs between development and conservation across the lion range to allocate conservation resources to effective landscape-scale planning.[38]

Okavango Delta, Botswana. © Colin Bell

Maasai Mara National Reserve, Kenya. © Jeffrey Wu

02

The most charismatic of cats

Understanding the individual personalities of lions could be the way to reduce their problematic behaviour before it begins.

Dr Natalia Borrego

I felt a rush of adrenaline as my Land Cruiser rolled to a stop next to the pride of lions. It was my first field season and the first time I was so close to lions in the wild. Their massive physique is much more imposing when you are sitting with the windows down without the usual safety afforded by zoo glass or metal barriers. The initial rush faded as it became clear that these lions had no interest in me. They had more important concerns, like taking comfort under the shade of a tree and napping.

It's been over a decade since I first set foot in Africa and became acquainted with these charismatic cats. Over the years, the rush has been replaced by familiarity. The time I spend with lions in the bush no longer feels foreign but more like a second home … a home worth preserving.

My interest in studying lions was first sparked by their fascinating social behaviour. Lions are one of the most remarkably social of all mammalian species.[1] Within the realm of social carnivores, they are the sole representative of the feline family. They are the only cat species where both males and females live in permanent, stable social groups based on prides (related groups of females) and coalitions (groups of males).

Association patterns among pride and coalition members are highly dynamic. Individuals are rarely found all together; instead, group members scatter into smaller subgroups throughout their territory. This social organisation is known as fission-fusion and allows lions to flexibly adapt their group sizes. They get to maintain the benefits of belonging to large social groups – such as cooperative resource defence – when 'fused' but can also 'fission' into smaller subgroups to reduce within-group competition when resources are limited or tempers are frayed. The fluidity of a fission-fusion society is a key feature of lion sociality and allows them to exist across a wide range of habitat types and under varied ecological conditions.

Contrary to the popular belief propagated in films and fables, lion societies have no king. Unlike many other social carnivores, lions do not form dominance hierarchies, meaning their behaviour is not governed by strict social rankings. Instead, they are egalitarian, making them 'one of nature's few true democrats'. For example, pride mates do not adhere to any apparent feeding hierarchy and all females have equal access to their pride's territory. In males, access to mates is not determined by social rank but rather on a first-come-first-served basis. This gregariousness lends itself to cooperation among group mates. Indeed, lions engage in an impressive variety of cooperative behaviours, including cooperative hunting, cooperative territorial defence and the communal raising of their young.

Lion cub after a feast in the Ndutu region of the Serengeti, Tanzania. © Phil and Helen McFadden

Tawana Camp, Moremi Game Reserve, Botswana.
© Martin Harvey

Their lives are complex. Individuals must navigate a rich social landscape while contending with the ecological challenges of life on the savanna. In other species (e.g. apes, elephants and dolphins), social complexity has been linked to intelligence; the social intelligence hypothesis is that social complexity bolsters the evolution of cognitive complexity. This was my impetus for studying lions. I wondered whether lions possess the same cognitive skills documented in other social species. I wanted to investigate the cognitive abilities that underlie their complex behaviours.

Lazy brutes or clever cats?

In 2009, I set out to explore these questions. By this time, science had moved well past a view of animals as machines incapable of thought or reason. However, investigations of animal cognition were still limited to a few select systems, like primates and cetaceans. Carnivores were grossly underrepresented in the scientific literature. Even more surprising, formal studies of cognition in the big cat family (e.g. lions, leopards, tigers and jaguars) were non-existent.

In those days I was often met with scepticism or incredulous amusement. What was the value of studying cognition in lions? They are only big, lazy brutes. Yet, determined to find answers, I forged ahead and designed experiments to test lion cognition. Worst-case scenario, I would confirm the majority opinion of lions as lazy brutes. Best-case scenario, I'd find that lions are clever cats and likely to rely on their smarts to navigate complex social and ecological landscapes. Spoiler: lions are incredibly clever cats.

I investigated lion cognition using a classic method in animal behaviour studies: the puzzle box. I designed a box to use with lions that was modelled on boxes used in other (smaller) species, like monkeys. The lion puzzle box had a door on the side that was held closed by a spring-loaded latch and could be opened by pulling on a rope attached to the latch. Once open, the box was large enough to accommodate an adult male lion reaching his head inside. To test lions' smarts, I baited the inside of the box with a desirable chunk of meat, closed the door and then left the box for a lion to attempt to solve.

I conducted these experiments in various zoos and sanctuaries throughout the United States and South Africa. I also performed a subset of experiments with wild lions to confirm that their performance did not differ from captive lions – there was no difference. Lions approached the puzzle-box challenge with gusto. Upon first encountering the box, they quickly set out to obtain the meat inside. Individuals displayed remarkable exploratory diversity and persistence. Often a lion would first approach the puzzle by clawing, biting and scratching the box. Some individuals appeared to get frustrated and would frantically dig under the box, creating a cloud of dirt and sand, before walking away for a bit… almost as if they were 'having a think' or calming down before approaching the puzzle anew.

Eventually, the majority of lions (76%) figured out how to open the box, demonstrating cognition associated with innovation and problem-solving.[2] Perhaps most interesting, lions were better at solving this task than their asocial relatives, tigers and leopards, lending credence to the hypothesis that sociality can bolster the evolution of intelligence. In subsequent experiments – and akin to findings in other social species – lions demonstrated cognition in the forms of learning, memory, repeated innovation and cooperative problem-solving.

In the time since my first experiments, the field of animal cognition has progressed. Carnivores are a much more well-studied group. Research has moved beyond simply describing or comparing cognition across species. We are now asking: how can our understanding of animal behaviour and cognition be applied to conservation issues?

No lion is an island

In tandem with captive lion research, I study the behaviour of wild lions. Through my years in the African bush I have become acquainted with the many threats facing wild populations, as well as the uphill battle and harsh realities of trying to conserve them. My experiences led me to expand upon purely academic questions and ask whether understanding lion behaviour could inform conservation practices.

Unfortunately, current conservation approaches largely overlook fundamental behavioural traits. The common practice of translocation is a revealing case study.

A key aim of lion conservation is developing tools to successfully mitigate human-lion conflict, with emphasis on the word 'successful'. With the human population expanding at a rapid pace, settlements are encroaching on protected areas. As a result, humans and lions encounter each other more and more frequently, generating conflicts. The primary source of conflict arises when lions kill livestock. In turn, humans kill lions, both in retaliation and pre-emptively. Lions that depredate livestock frequently make a habit of the behaviour and become what is termed 'problem lions'.

Solving the issue of a problem lion is often achieved by euthanasia or translocating the lion to a different area. Both tactics are challenging. I will focus on the latter tactic, translocation, as the issue of euthanasia is self-evidently at odds with conservation. In spite of this, in the absence of effective management techniques, euthanasia may be the preferable tactic. I will explain why.

On the surface, translocation appears to be a compassionate and straightforward solution. Simply move these lions from their current territory where they are causing problems to a different location in a protected area. This approach presents a win-win solution: a win for conservation in that the lions get to live, and a win for the people whose livestock are no longer under threat from dangerous predators. Unfortunately, this outcome is rarely the case, and the reality is much grimmer.

By translocating lions, we are ignoring what should be a key consideration in their management: their social behaviour. This oversight is highly tricky for a win-win scenario. Moving a problem lion means removing it from its social group and no lion is 'an island, entire of itself'. Removing a lion, in fact, has detrimental consequences both for the problem lion itself, and for the companions it leaves behind.

Lions depend on their group mates to help take down large prey, maintain their territory, raise cubs and defend them from infanticidal rival lions. Removing just a single male lion from its coalition weakens the coalition's ability to defend its pride. Coalitions of males compete for access to prides. Once a coalition gains residence with a pride, which is no small feat, it must defend the pride against rival males. If a coalition fails to defend its pride (which includes cubs they have sired) from takeover, the rival coalition ousts the residents and kills all the cubs. As we can see, altering the social structure of lion groups by removing problem individuals has important, potentially deadly, consequences for the lions left behind.

The translocated lion also faces an uphill battle on multiple fronts, a battle it often loses. Available habitat is an increasingly scarce commodity in Africa; space is limited. The options that do exist are likely already home to resident populations, which will attack and vigorously defend their resources from unfamiliar individuals. This means translocated lions are separated from their companions, moved into unfamiliar territory and must navigate this new landscape by themselves while attempting to avoid the already resident lions. Solitary lions already have a rough go of it and are rarely successful, even under the best of circumstances.

One of the first wild lions I studied, a female dear to my heart, was eventually deemed a problem lion. She was translocated into an area with a resident lion population. Although she fought a noble fight, she ultimately did not survive the translocation, thanks to a combination of unfamiliar territory, loss of her pride mates and attacks by the resident females.

Formal investigations of translocation outcomes confirm its ineffectiveness. The majority of translocated lions die; the most recent study reports a 76% mortality rate.[3] In addition, most translocated lions continue to kill livestock. Thus, translocation does nothing to alter the problematic behaviour, it merely transfers the challenge somewhere else.

Significantly, not all lions living in close proximity to humans become problem animals. Understanding why some lions exhibit problematic behaviours – such as killing livestock – while others do not, may be the key to preventing these behaviours from arising in the first place. Individual differences in behaviour imply that animals have 'personalities'. Personality was once a taboo word in my field, but we have now come to recognise that animals show significant inter-individual variation in behavioural traits.

Young lions on a kill on the edge of a wetland in Zakouma National Park, Chad.
© Michael Lorentz, Safarious

Unsurprisingly, an investigation led by my master's student (Victoria O'Connor) found personality differences among lions.[4] In other species like coyotes, personality traits such as shyness, boldness and wariness can alter individual responses to the same management tools and may signal an individual's potential predisposition to becoming a problem animal.[5]

Studies aiming to type individual lions in the wild behaviourally might allow us to identify potential problem individuals and the underlying causes of problematic behaviours before they become an issue. This approach would enable managers to mitigate conflict before it begins. By ignoring fundamental aspects of lion behaviour, our current management interventions seem more likely to fail. Yet if we alter our approaches, there is reason to hope.

Lion populations are rapidly declining with a notable exception: southern Africa, which provides a striking example of a conservation success story. In the 1900s, lion populations in South Africa were reduced to only three groups. Following a laudable reintroduction and management effort, South Africa is now one of the few countries where wild lion populations are increasing rather than declining. Lions were reintroduced to 40+ reserves in South Africa and rapidly repopulated. In fact, some South African reserves are struggling with a surplus of lions.[6]

However, this success story has a caveat. In order to protect their wildlife, South Africa's reserves are heavily managed and often fenced around the entire border. Fences serve an important purpose – securing habitat by separating lions from people, thereby significantly reducing human-lion conflict. But fencing is not a viable solution everywhere; conservation is not a one-size-fits-all answer. We must be flexible and innovative in our approaches. South Africa's example does offer an important lesson: lions are a resilient species and when we protect their habitat, they are capable of astounding recoveries from the brink of extirpation.

03

Saving Africa's lions

The African lion is one of the most iconic species in the world. Just look at all the lions in art, as statues, on logos, on flags. The door knocker on Number 10 is a lion's head and massive lion statues sit on Trafalgar Square.

Dr Pieter Kat

Add to the above list all the heraldic symbols of lions so favoured by European royalty. There are also lion statues in places like China, Indonesia and Cambodia, where no lion ever set paw. Elephants, rhinos, chimpanzees, gorillas, leopards, polar bears? Hardly any.

For humans, lions represent strength, protection, courage and loyalty, so where did it all go so wrong for these majestic cats to the point that they are now an endangered species? Here are some of the reasons.

Genetics

It may come as a surprise to some people to learn that there is actually no such animal as an African lion. That is the first thing that African conservation organisations and wildlife departments need to accept to ensure effective conservation programmes.

Sure, there are lions in East, southern, Central and West Africa, but a great amount of recent evidence has shown these are all genetically distinct populations.[1] Indian lions are rightly afforded subspecies status distinct from African lions. After all, they're a long way from Africa. Unexpectedly, it has now been shown that western and central African lions are more closely related to Indian lions than their supposed genetic relatives in eastern and southern Africa.

This could have been anticipated. North African lions were extirpated in the wild. There might be some surviving in zoos in Morocco and Europe (crossbred with eastern and southern African lions in many cases), but what was called the Barbary lion is officially extinct.

But long before the pharaohs, those Barbary lions and others further east in Egypt most likely found an exit route from Africa into the Middle East, Central Asia across Iraq and Iran, and deep into India. Those same North African lions travelled to Europe. Cave paintings in the Chauvet Cave in France record their presence about 30,000 years ago. Of course, some stayed in West and Central Africa. The most likely original cause for this genetic divergence was the Sahara Desert and the Sahel, isolating north/south lion populations, exacerbated by Pleistocene glaciations that dried Africa even more than it is now.

While most of the continent is experiencing a decline in lion populations, southern Africa generally stands out as a stronghold. In many areas, the success of populations is attributed to fenced protected areas, which unfortunately have the unintended consequence of hindering the natural migration of wildlife. To address this, artificial measures are implemented to maintain genetic diversity. Typically, individuals are carefully chosen for removal and relocation to a new reserve when they reach an age at which they would naturally disperse in an open system. The process involves immobilising the selected lions, after which they are either loaded into crates for transportation by road or, in certain cases where the journey is too extensive or hazardous, they are kept immobilised and transported by air to their new habitat.

The genetically different eastern and southern African lions stayed put. Their own ranges might have expanded and contracted over the millennia – and certainly, eastern and southern African lions also show genetic differences – but, genetically speaking, this is more likely to be a 'cline' as there are populations in Mozambique and southern Tanzania connecting those in, say, Uganda with South Africa.

What remains of the western and Central African lions is small and decreasing. There used to be a small stronghold population in a cross-border protected area in Benin, Burkina Faso and Niger, but with ongoing civil strife, meat poaching, and hardly anything that can be called a 'wildlife department', it's doubtful those lion populations can still be considered viable. These lions need urgent conservation attention and much funding, but how can this be achieved in nations where various militias and perhaps national armies are competing for territory?

East African lion populations are similarly declining, but for different reasons. According to the International Union for the Conservation of Nature (IUCN), southern African lions are on the increase. However, this is hard to verify, as there have been no wild population counts across Africa. The count is based on lion populations in South Africa, most behind fences – the traditional South African conservation model these days.

If there is to be a concerted effort to conserve lions across Africa, different means and models must be applied, tailored to each range state with lions and ranked in order of urgency, and in recognition of the need to maintain as full a range of African lion genetic diversity as possible. Those nations will have to fund (or request independent funding) to nail down what lion numbers remain.

Determining what remains

African lion range states seem unconcerned about achieving at least a minimally acceptable international standard of lion population counts. Perhaps there has been a limited census in a few protected areas here and there, though they are difficult, time-consuming and expensive.

Of course they are, especially when such counts cannot just be a once-off and should be repeated after regular intervals; and then optimally matched with images of individual lions counted and identified before in a specific area to determine overlaps or new individuals.

Even lion researchers located in areas with significant lion populations rarely conduct regular comprehensive lion surveys. Not in Hwange, Okavango, Chobe, Serengeti, Ruaha or Luangwa and definitely not in the far corners of the Selous. These researchers should be at the forefront of providing the best information available.

I estimated in 2020 – on the basis of information from a great diversity of sources – that there were perhaps around 10,000 lions remaining in the wild in Africa. This number was met with great opposition, not from the African lion range state wildlife departments, but from lion trophy hunters and their bedfellows. It is a better estimate than 'maybe 20,000' provided by the IUCN, built on extrapolations of 'well-monitored' populations skewed by lions in fenced private parks in South Africa.

It should be noted that South Africa's largest national park, Kruger, has not conducted a lion population survey with any scientific rigour for over 13 years and those estimates were based on 700 lions seen (not individually identified, some could be double counts) and a population of about 1,600 extrapolated. However, Kruger does seem more interested in conducting a park-wide survey than many other protected areas in African range states.

Increasing lion protection

We have a real problem understanding lion numbers outside of protected areas for two reasons. In these areas, nobody is counting lions in any replicable way. At issue is that it's not known how many lions occurring within protected areas could also be the lions seasonally occurring outside of them. I am not talking about fenced areas that occur around private and nationally protected areas in South Africa, but elsewhere where national park borders are minimally fenced, or not at all.

> **If you were guaranteed *not* to see a lion on a visit to your African safari nation of choice, would you still visit? The clear answer was a resounding 'no'.**

Despite some fence-building by Kenya – as well as some other African lion range states – there are still few 'hard' borders between a nationally protected area and, let's say, community-owned lands, commercial hunting areas, photographic tourism areas, areas leased out to operators by governments and many more such categories of land ownership or lease arrangements.

Take, for example, Amboseli National Park in Kenya, one of the most famous national parks, which attracts hundreds of thousands of tourists annually. The survival of its biodiversity and individual animal populations is greatly dependent on the availability of Maasai tribal lands directly bordering the park. Amboseli is not a stand-alone ecosystem. Maasai communities are expected to allow wildlife to graze on their lands and also accept seasonal incursions of predators.

While the Kenya Wildlife Service consents to this formula, it remains reluctant to share the significant gate fees with surrounding communities. Neither do the highly profitable tourism companies operating under big brands, which put little into surrounding Maasai communities.

If wildlife populations are going to thrive in Kenya – where about 10% of that nation's GDP comes from tourism and an even bigger percentage contributes to its foreign income earnings – this formula needs to change. I asked a simple question many years ago: If you were guaranteed *not* to see a lion on a visit to your African safari nation of choice, would you still visit? The clear answer was a resounding 'no'.

Causes of African lion declines

The causes of African lion declines are legion. One is the loss of what has been called the 'lion estate', areas where lions can roam. Loss of lion natural prey owing to rampant wildlife meat poaching across Africa is another. Without natural prey, lions inevitably turn to livestock. Conflict with humans and livestock results, and while there are any number of schemes available to mitigate and prevent such conflict, I see only limited effective measures applied by governments or non-government organisations (NGOs).

Trophy hunting is another issue. It kills the biggest and best males in lion populations and is accepted by a diversity of African nations as 'sustainable offtake' – without proof. The consequences of trophy hunting have been effectively suppressed by hunting operators, governments and global associations who claim sustainability while never having conducted population surveys and indeed risk-benefit analyses of lion trophy hunting. Then there's the threat that lion populations in nationally protected areas are becoming ever more isolated as migration corridors increasingly disappear.

If there's a will to conserve what remains of Africa's lions, there have to be significant interventions. The status quo can no longer be tolerated and will drive lions ever further towards the cliff edge.

Unless effective and broad-scale programmes are implemented by local and international NGOs operating in Africa, supported and augmented by local governments, lions will continue to decline. Conservation of large predators has been proven difficult across continents, but surely positive lessons learnt can be applied to African lions. In 1989, Douglas Adams and Mark Carwardine wrote a book called *Last Chance to See*.[2] Over 30 years later, their warning has still not been heeded for African lions.

Failing this, in the worryingly near future, you will have to buy an entry pass to your local zoo to be given a last glimpse of a globally iconic species that disturbingly few took courageous and innovative efforts to conserve.

04

Genetic variability in lions

We frequently focus on preserving species or preserving ecosystems, the places these species inhabit. But this is not enough. There's a third pillar of biodiversity, one more fundamental than the others: genetic variability.

Dr Laura Bertola

How do we assess genetic variability within a wide-ranging species like the lion and what can it tell us about the past, the present and the future of the species? How do we monitor changes in this variability and how can we use this to inform management and policy? We should not focus on saving 'just' lions, we need to focus on saving lions in their full diversity.

It is widely known that we are losing biodiversity at an alarming rate. But often, these statements focus on the loss of species as a result of extinctions. The much larger chunk of diversity we're losing, however, is on the intraspecific scale, encompassing the genetic variability within species.[1] For species that have a wide distribution, such as the lion, this is especially relevant. Populations in various parts of the range reflect different evolutionary lineages, formed by processes over hundreds of thousands of years. This means that if lions go extinct in parts of their range, it potentially means we are losing an entire branch of the lion's evolutionary tree. It has been widely shown that loss of genetic diversity increases the risk of extinction, and is informative about population persistence, adaptation and resilience, making it directly relevant for conservation.

Intraspecific variability can best be seen in a spatial context. It's not only relevant to know how diverse populations are, but also how different genetic lineages are distributed across the species' range, and how differentiated they are from each other. These patterns give us an insight into historical connectivity, and possibly even what the landscape must have looked like in times of the lions' ancestors.

Once, lions were the mammal with the widest distribution, apart from humans. Subspecies that are now extinct ranged throughout Africa, into Europe and Asia and across the Bering Strait into North America.[2] These old species and subspecies went extinct in the late Pleistocene, around 14,000 years ago.[3] What we now know as modern lions still had a continuous range, stretching out of Africa into Asia and southern Europe.[4] In Europe, lions were likely driven to extinction between 3000 and 1000 BC; and they disappeared in the Near East, Arabian Peninsula, Transcaucasia and the north of Afghanistan around the 12th and 13th centuries AD. Populations in the North African countries and the Middle East were lost between the end of the 19th century and the first part of the 20th century, largely as a result of human persecution.[5] Lions are now restricted to an increasingly

A normal-coloured lion cub and a white lion cub born from the same litter in the Timbavati Private Nature Reserve, Greater Kruger area, South Africa. © Chad Cocking

fragmented range throughout sub-Saharan Africa, with the exception of dense rainforests and dry deserts. Compared to the 1960s, it is estimated only 25% of suitable savanna habitat remains, and the decline is ongoing.[6] Parallel to the decline in suitable habitat is the decline in lion numbers: an estimated 75% in just five decades.[7] There is only one population left outside of Africa, in the Gir Forest in Gujarat State, India. Because of the extinction of all intermediate populations historically connecting these Indian lions to African populations, the Gir Forest lions now face extreme isolation.

Until recently, lions were grouped into an Asian subspecies (*Panthera leo persica*), with the Gir Forest lions being the sole representatives, and an African subspecies (*Panthera leo leo*), covering all other populations. At first glance, this could make sense, since the Indian population distinctly stands out. However, genetic studies have shown this is not in line with the evolutionary history of the lion and does not reflect well the distribution of genetic variability in the species.[8]

Despite the local loss of connectivity through recent human-driven habitat conversion,[9] lion populations were broadly connected from West Africa, throughout Central Africa, into East Africa and south into southern Africa. However, when looking at their genetic makeup, we observe a clear divergence within Africa. Lions in East and southern Africa look like two closely related lineages, while lions in West and Central Africa comprise another lineage, strongly diverged from the ones in East and southern Africa.[10] When examining the evolutionary relationships between these lineages, it becomes clear that the Indian lions are similar to African lions in West and Central Africa. In other words, West and Central African lions are more closely related to the Indian population than they are to their counterparts in East and southern Africa.

Based on this new insight, and in combination with other lines of evidence, the taxonomy of lions has been rewritten. Now we distinguish a northern subspecies (*Panthera leo leo*), grouping lions in West

Dawn in Pendjari, Benin, in an area of extensive three-metre-high grasses. Walking along the road, we found a footprint trail with occasional diversions off to either side before returning a little further up the road. When we came to the end, the obvious decision was just to sit and wait. After half an hour or so of sitting quietly, we were rewarded with a brief glimpse of the rarely seen West African race.

and Central Africa with the ones in India, and a southern subspecies (*Panthera leo melanochaita*), covering populations in East and southern Africa.[11] This new taxonomy is in line with evolutionary history and, more important, affects how we prioritise conservation action. After all, if we cannot manage to conserve lions in the extremely fragmented and small populations still existing in West and Central Africa – where local extinctions are common and where remaining populations are under huge pressure – we are not 'just' losing lions, we're losing an entire branch of the lion evolutionary tree.[12]

Although insight into the spatial distribution of genetic variability in itself is already extremely helpful for informing conservation actions, it is crucial to have some understanding about the processes that gave rise to the patterns. Why did lions within Africa diverge, especially since the northern and southern subspecies actually overlap in Ethiopia? Here, lions of both subspecies not only meet each other, but also interbreed, leading to a hybridisation zone that becomes visible only when studying the genomes of these populations.[13]

In earlier studies, it was hypothesised that the location of the Rift Valley, a large geographical feature running from Ethiopia to Mozambique, may form a barrier to dispersal.[14] However, lions have been observed to cross rough terrain in the Rift Valley, and populations in Ethiopia show a mixed ancestry between both subspecies. If contemporary barriers do not explain the observed patterns, it may be that there have been barriers in the past that have since disappeared. Indeed, climatological data – as well as pollen studies, which have reconstructed and modelled the extent of the Central African rainforest – shows that the borders of the forest have changed extensively during the past 300,000 years.

Typically, in this part of the world, deserts expanded in a north–south fashion, whereas forests there expanded along an east–west axis. It could be that these expansions, following oscillating climate changes, have periodically reduced suitable lion habitats to small and isolated patches.[15] Lions may have survived in these local refugia and diverged from other isolated populations during periods of isolation. When environmental conditions became more favourable, these populations may have expanded their range and come into contact with other such diverged populations, locally leading to a hybridisation zone.

We can use divergences observed from genetic data to date the timing of the split between two populations; and the timing of such splits seems to coincide with changes in the climate that likely impacted the range of suitable lion habitat.[16]

Genetic variability

Genetic variability can be measured on different scales, ranging from divergence between populations to diversity within populations or even within individuals. Therefore, its loss occurs on different scales. On a larger geographic scale, we lose populations as a result of local extirpations, resulting in the loss of uniquely diverged genetic lineages, as outlined above. On a smaller geographic scale, we lose diversity within populations, as they become confined to smaller and more isolated patches of habitat. As populations shrink and opportunities to exchange genetic variability with neighbouring populations are reduced because of isolation, diversity will erode.

Partially, because of a force we call 'genetic drift', the change in genetic composition results from chance events, which is particularly strong in small populations. The second reason small and isolated populations lose diversity is because all individuals will gradually become related to each other. And when related individuals start breeding, chances increase that deleterious mutations leading to heritable diseases become expressed. This results in a decline in fitness and is termed 'inbreeding depression'.

Small and isolated populations can enter the 'extinction vortex' in which they lose diversity through demographic and genetic processes, making the population even smaller and further accelerating the loss of diversity. Although it's not easy to restore lost diversity, conservation management can intervene by adding new genetic variability to populations through, for example, translocations; this would safeguard populations that would barely stand a chance without intervention.

The true interpretation of genetic patterns without historical context is challenging. Imagine that we analyse genetic data for several lion populations in a given country and the results show there is strong genetic population structure. How do we translate this finding into a conservation recommendation? Having some insight into how the landscape has looked in the past and how these lion populations have been distributed in the past may help.

If there has been high connectivity in recent history, then the current population structure may be the result of populations becoming small and isolated in pockets of suitable habitat; and a possible recommendation could be to restore past connectivity. This could be done through ecological restoration – actually creating new suitable habitat, which links up these isolated pockets – or by transporting individuals from one place to another, hereby mimicking natural gene flow.

However, if there are indications that there has been no recent historical connectivity, the population structure we observe now may be something we would like to maintain, and the isolated populations should be managed separately. Hence, we would advise against connecting these pockets of habitat or moving animals around. After all, these diverged populations may reflect distinct evolutionary branches on the tree, which may suggest adaptations to the local environment.

Using the data

How is genetic data currently being used in lion conservation? The revision of the taxonomy, highlighting the diverged position of West and Central African lions, has likely been helpful to prioritise these populations and focus conservation funding on this part of the range. It has led to a discussion within the zoo community, which currently has a breeding programme only for African and for Asiatic lions. As such, the current taxonomy is not reflected in their management. Ideally, the ex situ population would reflect the variability in the wild and could function to safeguard the particular diversity that we may not be able to conserve in the wild.

To assess the genetic variability of ex situ lions, ~70 lions from European zoos (associated with the European Association of Zoos and Aquaria, EAZA) were genotyped. Results show that, although zoo lions typically show a stronger mix of diverse ancestries, we can still distinguish different genetic lineages or traces thereof in the zoo population.[17] This information is now being used in discussions about breeding programmes, the capacity to manage these and the prioritisation of genetic lineages that are under extreme pressure in the wild.

The genetic background of populations – such as the assignment to a specific genetic lineage – is important for management interventions such as translocations. During translocations, lions can be released in an area where lions are no longer around (reintroduction) or in an area with a resident population (reinforcement). The latter is typically done if a population is showing signs of inbreeding depression.[18] The rationale is that new genetic information is hereby added, leading to increased fitness in the offspring and lifting the population out of the extinction vortex.

But scale is crucial. On a local scale, adding unrelated individuals to a resident population may increase diversity, similar to the way natural immigration would introduce new variability. However, on a larger geographic scale, mixing between strongly diverged populations may not be advisable. The hybridisation of strongly diverged lineages can result in genetic incompatibilities and therefore reduced fitness: 'outbreeding depression'.

This is difficult to monitor in the wild since fitness data is hard to obtain and loss of fitness may show up only after two or three generations. Our current understanding is that outbreeding depression is less of a risk than inbreeding depression, but reliable data from wildlife is still sparse. It highlights the need for long-term monitoring after a management intervention to establish the outcome of the intervention and inform further steps.

In addition to a potential threat of outbreeding depression, a consequence could be that in mixing, lineages homogenise and boundaries between lineages become blurry. This sometimes happens naturally, when populations expand and hybridise with neighbouring lineages. Yet, if human intervention leads to homogenisation, it probably means suboptimal populations were selected for the translocation and, as a result, variability is lost. This is particularly relevant in the context of lions since data from the Convention on International Trade in Endangered Species of Wild Fauna and Flora (CITES) – which stores all permit information since the 1970s – shows that the vast majority of translocated lions come from South and southern Africa.[19]

South African lions have already been shipped as far as Malawi and Rwanda for reintroduction projects. In these cases, lions from East Africa would have been more suitable candidates because they would have been more closely related to the populations that roamed those areas originally. Of course, we must acknowledge that managers and policymakers involved in the decision-making may not have access to the relevant scientific literature and that many non-genetic considerations also need to be considered during the decision-making process.[20] Sometimes there is pressure to make a quick decision, which does not allow for long deliberation or genetic testing of animals.

In a recent study we show that, based on information we have gathered over the past years, it is feasible to assign all contemporary lion populations (with varying degrees of certainty) to a specific genetic lineage.[21] On this information, we can identify which populations are suitable for translocations and which are not.[22] We hope this can serve as guidance and support for managers in their decision-making, providing them with a range of options rather than restrictions.

To successfully integrate genetics into conservation management and policy, we have to ensure that both the data and techniques to generate this data become accessible and broadly applicable. Cheap and flexible methods that do not require elaborate laboratory equipment are needed to assess the current state of genetic variability, and for monitoring and detecting changes over time, so that management can adjust accordingly. Genetic panels can provide an easy and cost-effective solution for this.

The design of the panel depends heavily on available samples and requires testing and validation on a range of sample types. However, once a panel is available, this may greatly facilitate ongoing and future activities. For lions, such a panel now exists and provides reliable data even from non-invasively collected samples, such as hairs, scat and saliva.[23] This opens up possibilities, such as asking poaching patrols and rangers to opportunistically collect samples when they encounter them.

Saving lions is not enough; we must also save the diversity within lion populations.

The panel has proved to be powerful enough to recognise individual lions and can be applied to a wide variety of conservation-relevant questions. For example, the genetic panel can be used to confirm lion presence in the area, estimate population sizes (by using a capture-mark-recapture framework), and estimate connectivity, population structure, overall variability and relatedness. In addition to genetic data that's relevant for population management, this type of data also has forensic value. The panel performs well on low-quality samples, which are usually the only material available after confiscation of often processed goods.

Genetic analysis can be used to confirm the species identity of a sample and perhaps the source population. This information can then be used to increase security measures locally or follow criminal trade routes internationally.[24] If baseline data exists in sufficient resolution, it may be feasible to recognise the poached individual. Individual identification can be more relevant because the ability to determine the number of individuals illegally killed may affect how seriously the offence is deemed and will affect conviction.

Conserving biodiversity within species is important because it is directly linked to resilience and persistence and contains the adaptive potential of the species. This enables evolutionary responses to changing environmental challenges. As Ceballos, G. et al, write: 'Dwindling population sizes and range shrinkages amount to a massive anthropogenic erosion of biodiversity and of the ecosystem services essential to civilisation'.[25] It highlights the need to monitor, manage and preserve diversity below the species level.

Saving lions is not enough; we must also save the diversity within lion populations.

05

The poison problem

The targeted poisoning of lions has contributed to the decline in the African lion population over the last 25 years and the extirpation of the species over much of its historical range.

André Botha

On 5 December 2015, shocking news broke of the poisoning of the famous Marsh Pride of lions in the Maasai Mara Game Reserve in Kenya. This tragic event occurred after the pride had fed on the carcass of a cow laced with carbofuran by herdsmen who had been illegally grazing their cattle in this protected area.

For a short while, the poisoning of African lions grabbed global headlines and the world followed the efforts of conservationists to save the animals. Three of the pride were confirmed dead, while several individuals disappeared and were never accounted for. Since this headline-making incident, at least 160 other lions from nine countries (Figure 1) are known to have been poisoned across sub-Saharan Africa.

Because of their predatory nature and reputation as killers of livestock and, occasionally, of people, lions are often the target of poisoning. Baits are used either in response to the loss of livestock or human lives, or as a preventative measure when people feel threatened by lions in their vicinity. Lions are also targeted for reasons other than human-wildlife conflict and can become the unintentional victims of baits placed out to kill other target species regarded by some as problem animals, such as leopards, hyenas and vultures.[1]

Historically, most poisons were derived from natural sources such as plants, animals and minerals. However, the manufacture of a range of agrochemicals for mass food production and pest management changed all that. Poisoners now have access to a wide range of potentially harmful pesticides that are easy to use and, in many countries, poorly regulated.

The lethal impact of these chemicals is far greater because of the wide range of species susceptible to them and their potential to poison non-target animals. In sufficient quantities on a baited carcass, the acute toxicity of substances such as methomyl, aldicarb and carbofuran can indiscriminately wipe out an entire lion pride and a range of other species too, both through primary and secondary poisoning. What's more, these are substances that can persist in the environment for a considerable time without undergoing a reduction in their toxicity.

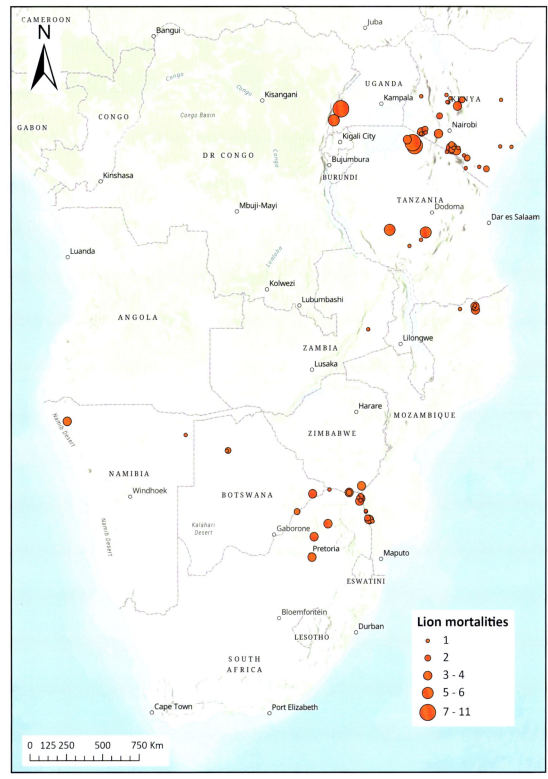

Figure 1 Wild lion mortalities in Africa from poisoning between 2014 and 2022.

Victims of their status

Lions feature prominently as a symbol associated with a wide range of beliefs in many cultures in Africa and beyond. Some linked rituals require lion body parts, and poison is the method of their harvesting. The body parts can be for local use, but also for use beyond the lions' range. Between 2008 and 2020, South Africa issued annual permits for the export to the Far East of lion bones – 1,500 carcasses in 2018 – for use, among other things, in the manufacture of tiger bone wine.[2] This process was stopped in 2021, but the demand for lion parts has not diminished.[3]

Lions in the wild are increasingly the target of poisoning in trade for ritual use. The killing of four lions in the northern Kruger National Park, South Africa, in September 2015 for this purpose[4] – followed by several more incidents in the same area over the last eight years – was largely driven by the demand for lion bones and other body parts. Similar incidents have been recorded in the Limpopo National Park,[5] Niassa Special Reserve[6] in Mozambique, and Queen Elizabeth National Park[7] in Uganda, among other areas. Finding mutilated lion carcasses with body parts such as the head, feet and claws, skin and certain organs removed after the animals have been poisoned is no longer an uncommon phenomenon to those who respond to such incidents.

Counting the cost

The African Wildlife Poisoning Database is an initiative of the IUCN Species Survival Commission Vulture Specialist Group with the support of the Endangered Wildlife Trust (EWT) and The Peregrine Fund.[8] It collects and collates information on all known wildlife poisoning incidents in Africa to assess the scale of the threat and identify wildlife poisoning hotspots.

Figure 1 provides a summary from this database of the known wild lion mortalities from poisoning

Poisoned lions (above left) and vultures (above right) are destroyed to prevent secondary poisoning of scavengers.

Poisoned vultures being piled together for incineration, Mbashene, southern Mozambique.

between 2014 and 2022. These figures exclude the poisoning of captive lions, which is a further issue.

Information collected by the EWT indicates that between June 2016 and July 2018, there were at least 21 reported incidents of captive lions poisoned leading to the death of 60 animals in South Africa. The chemicals used by the poachers in all of these incidents were carbamate or organophosphate agrochemicals, similar to the chemicals used in the poisoning of wild lions.

Targeted poisoning of lions has contributed to the extirpation of the species over much of its historical range. Protected areas are by no means effective safe havens and animals are frequently targeted and poisoned inside parks and nature reserves. A significant number of the incidents reflected on the map in Figure 1 occurred either inside or in the immediate vicinity of a range of protected or game management areas.

One example is the lions that were reintroduced to the Liwonde National Park in Malawi during the 20th century on at least four occasions. With each reintroduction, the animals were poisoned and eliminated from the park. More recently, in November 2022, a pack of 18 African wild dogs was killed in this park in a single poisoning incident, effectively wiping out the entire reintroduced population.[9]

Given the size of many African protected areas, incidents like this are difficult to prevent and even more difficult to detect. As a result, poisoning scenes may be discovered only weeks and months after the event, making investigation and successful prosecution a rarity. Even when perpetrators are identified, charged and convicted, penalties are often insufficient to be a deterrent. This is attributable to the justice system's lack of appreciation of the impact of these incidents on wildlife populations as well as the risk the poisons pose to the health of human communities.

The African Wildlife Poisoning Database currently contains more than 1,500 wildlife poisoning incidents reflecting the deaths of more than 44,000 individual animals. Lions represent the highest number of mammal mortalities, with African elephants a close second. However, the group of animals that far outnumbers any other on the database is vultures, comprising more than 15,000 or one-third of recorded mortalities. These figures likely represent a mere fraction of the actual mortalities from poisoning as most incidents are either not detected or not reported.

The secondary poisoning of other species when lions are targeted is also a significant threat. Species impacted include scavengers such as hyenas, jackals and, especially, vultures, as well as invertebrates such as dung beetles and flies. Poisoned insects can attract and poison insectivorous animals including a wide range of birds and mammals such as the African civet. The secondary consequences of poisoning are thus far greater than they are currently understood and quantified.

There are many examples of this ripple effect, but two of the more publicised events are worth sharing. An incident of retaliatory poisoning on the outskirts of the Ruaha National Park, Tanzania, in May 2016 resulted in the killing of a single male lion thought to be responsible for several incidents of livestock lost by herders in the area.[10] The carcass of the lion was fed on by a range of avian and mammalian scavengers. This resulted in the deaths of 56 critically endangered vultures: 55 African white-backed and one hooded. In addition, two Bateleurs, a tawny eagle, a black-backed jackal and a spotted hyena were found dead after feeding on the carcass.

Another incident, this time in southern Zimbabwe in 2016, resulted in the death of 41 African white-backed vultures after they scavenged the carcass of a poisoned lioness suspected of killing cattle.[11] This was followed by an incident in February 2018 when six lions and 74 vultures were poisoned.[12]

Working for lions

There are many organisations across the African lion's range working towards their conservation in the wild through a variety of projects to reduce human-wildlife conflict and address other threats such as persecution, snaring and loss of habitat. Reducing the impact of poisoning of lions and other wildlife has been a focus of organisations such as the Endangered Wildlife Trust (EWT),[13] The Peregrine Fund (TPF)[14] and others working with a range of partners across Africa to create greater awareness of the risks of the indiscriminate use of pesticides to kill lions and other wildlife.

> It will be a sad day when the roar of the lion is no longer heard over Africa's wild places and when there are no more vultures in its skies.

The EWT and TPF have conducted training with more than 7,000 rangers, police, veterinarians and other stakeholders in 16 countries in sub-Saharan Africa. This focuses on enabling individuals to confidently identify wildlife poisoning of a range of species and to initiate rapid response strategies to secure such scenes. This includes how to conduct proper investigations, collect evidence and, most importantly, decontaminate sites to eliminate the threat of further poisoning. There is also a focus on mentoring more trainers to expand the network of individuals in relevant countries.

These efforts are showing good results. Although arrests and successful prosecutions for wildlife poisoning remain few and far between, there are signs of improvement, with greater awareness among conservationists, law enforcement, the judiciary and the general public about the impact and health risks associated with this practice. The sentencing of two perpetrators in September 2022 to 17-year jail terms following the poisoning of six lions in Queen Elizabeth National Park, Uganda, is an excellent example of how all of these elements contributed to a successful conviction and appropriate sentence.[15] However, many countries in Africa are still challenged by lack of capacity to undertake forensic and toxicological analysis. Both are key elements to support the successful investigation, identification and conviction of perpetrators.

Many animals can be saved if still alive when found, provided that skilled teams with the necessary equipment and training are rapidly mobilised. In the Marsh Pride incident in the Maasai Mara in 2015, the survival of eight members of this famous pride was entirely thanks to the quick and effective intervention by veterinarians and other support staff on the reserve. In another incident in South Luangwa National Park in Zambia in September 2016, the Zambian Department of National Parks and Wildlife collaborated with Conservation South Luangwa and the Zambian Carnivore Programme to successfully treat 18 lions in the field, which fed on a poisoned elephant carcass.[16] Expanding the capacity to treat poisoned wildlife in key lion strongholds can support conservation efforts by reducing the impact of poisoning on overall population numbers.

Much needs to be done to change misconceptions and to enable effective and continued conservation action to address the threat of poisoning. A multi-layered approach is essential. This should include:

- Banning and better control of substances most frequently used in the poisoning of wildlife and the environment and which could impact human health.
- Improving and promulgating appropriate legislation and monitoring its efficient implementation and enforcement.
- Supporting law enforcement through improved investigative and forensic capacity across the range.
- Addressing the illegal harvesting and trade in lion and other wildlife products through demand reduction initiatives and engagement with buyer markets. These are often unaware of the impact of their use of these products on the species involved.
- Continuing and implementing more innovative approaches to reduce the prevalence of human-wildlife conflict caused by lions and other wildlife on human livelihoods.

Should the current trends in the poisoning of wildlife continue, there is a real possibility that vultures, lions and other species may be eliminated from much of their current range in the next 30 years. Combined with other threats that impact wildlife, the extinction of some that are already teetering on the brink is a stark reality.

It will be a sad day when the roar of the lion is no longer heard over Africa's wild places and when there are no more vultures in its skies.

Identifying lions

Then ...

In the 1970s, Kruger National Park scientists and researchers experimented with different identification processes to help them to recognise individual lions. Coloured and numbered ear tags were tried, but they fell off after about six months. They settled on a branding scar created by a red-hot iron that was manually applied to the rump or the shoulder of an anaesthetised lion. After two weeks, a permanent scar developed in the shape of the branding iron that helped the authorities with their identifications. This black-and-white photo from Dr Butch Smuts' book *Lion* shows him branding a lioness. In the 1980s the Savuti sector of Chobe National Park was at its peak as the lion capital of southern Africa. Researchers started branding lions and then applying radio collars on a number of the region's key individuals, but this detracted from the feeling of remoteness and wilderness. Scientists and researchers have now learnt to identify lions through their markings, scars, whisker patterns and notches, removing the need to brand or collar.

Dr Butch Smuts' book *Lion* about his experiences and time in the Kruger was published over 40 years ago.

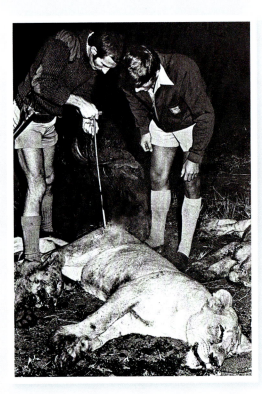

Now ...

This is how we identify lions today, taking notes on their features such as whiskers, whisker spot patterns, noses, notches on ears and noting the chips as well as the wear and tear on their teeth. Scars and scratches aren't often included in their identification unless the scar is large and permanent. Radio collars are still needed in certain areas where lions in unfenced wildlife areas do come into contact with livestock and people. In these instances, the radio collars are an important management and advance warning tool to keep lions, humans and livestock apart, helping to minimise or even stop retaliatory killings.

A lion's nose colour changes with age: a younger lion's nose is pink and by six years old it starts to turn a mottled dark colour. By the time it is eight years its nose is fully black. A male's mane will also change over time, so can often be used as an aid to help estimate the individual's age.

© Bobby-Jo Vial

Radio-collar tracking of lions aids research and provides early warnings to rural communities living alongside unfenced parks, enhancing conservation and safety.

Conservationists record identifying features of individual lions, such as whisker spot patterns, nose colour, scars, and wear and tear on teeth

© Dana Allen

Disease control
Understanding distemper

Back in 1994, 30% of the lions living in Serengeti National Park died from an outbreak of distemper, which came as quite a shock to park managers. Until then it was believed that cats could not catch the disease, which is normally associated with dogs. Distemper is a viral disease caused by the canine distemper virus (CDV). It has been known to infect dogs, coyotes, foxes, pandas and a wide variety of other animals, including big cats. Distemper is spread from unvaccinated infected dogs to wildlife and between infected wildlife. The domesticated dog seems the primary host but other wildlife species also act as a reservoir.

There is no carrier; distemper is spread only by infected animals themselves, which may survive the duration of the disease – typically up to a few months – or die from it. The virus survives only 20 minutes outside of the body and is spread through saliva, faeces or urine, though mothers can pass it on to offspring during pregnancy.

After being infected, symptoms can start from within two weeks to a few months. If an animal survives, it is then clear of the virus and develops lifelong immunity. Inversely, lion populations that have not been exposed to distemper for a long time may be at greater risk of high mortality when exposed.

The virus replicates in many different cells in the body, so symptoms are broad and include gastrointestinal, neurological, lung, skin and eye conditions. Prevention in dogs is easy with vaccination, though a series of treatments is needed. However, whether this vaccine works with wildlife is still under debate. One must simply remain supportive with fluids, antibiotics and seizure control.

Distemper is definitely more likely to infect young and immunocompromised animals, so the less healthy the lion population (stress, overcrowding, undernutrition, other diseases, inbreeding, etc.), the more chance there is of a distemper outbreak, though a healthy population is not immune.

Total prevention is not possible, but game reserves can minimise risks by ensuring that imported lion populations are healthy, that they have had no contact with unvaccinated dogs and that their population sizes, genetics and food sources are optimum.

06

Lions of the north

THE SAHEL

Lions are hard to count even in the best of places. Off the tourist track where they are rare and skittish and where infrastructure is lacking, there are no accurate numbers. A common estimate of their numbers is around 20,000 lions in Africa, of which only around 1,000 belong to the northern subspecies, *Panthera leo leo*.[1]

Prof Hans Bauer

This subspecies, in which I specialise, is distributed across the greater Sudan-Sahel belt from the Atlantic coast off Senegal in the west to the Ethiopian Highlands in the east.[2] It is bounded by the Sahara Desert in the north and the Congo rainforest in the south. Confidence intervals are wide, and we cannot exclude the possibility that the numbers are half or double 1,000; but, either way, the species remains extremely rare.

Though lions are not doing badly in southern Africa, where many viable populations have been stable or are even growing, in the rest of the continent their decline is ongoing and serious. In West Africa, the lion is listed as regionally Critically Endangered on the Red List of the International Union for Conservation of Nature (IUCN), and in East and Central Africa, they are classified as Regionally Endangered.[3] This decline started many decades ago, but with most of the lion conservation community working in southern Africa, it went largely unnoticed until 2001.

That year, I came back from Cameroon to the Netherlands to write my dissertation at Leiden University. In Cameroon, my communication had been by fax; during this time the rest of the world had rapidly embraced the internet. With new communication opportunities, I soon found people who had observed declines in many places, and we realised that we needed to create a network of interested parties.

In June 2001, we held a meeting in Limbe, Cameroon, with people from different countries in West and Central Africa. While the information we gathered there was, in retrospect, incomplete and imprecise, the insight was crystal clear: lions were not – and are still not – doing well. Our report was summarised in *New Scientist*[4] and picked up by the academic journal *Science*.[5] Within weeks I was invited to become a member of the IUCN Cat Specialist Group and now, two decades later, we have an IUCN African Lion Database that links the efforts of many lion conservationists involved in monitoring lion status across the continent. This is done in collaboration with both government and non-government organisations.

The general public is insufficiently aware of the situation, but the conservation community knows it very well: lions have become 'conservation

Niokolo-Koba National Park in Senegal was originally started to protect the western roan antelope in 1954, which were seemingly abundant throughout the reserve. The reserve is 913,000 hectares in size, comprises typical Sudano-Guinea savanna and houses some unique ecosystems and mammal species, including critical populations of African painted dogs, western Lord Derby eland and West African lions. NKNP is one of the last remaining strongholds of the critically endangered west African lions, conserving around 10–15% of the continent's remaining population.

dependent' and their survival as purely wild animals can no longer be taken for granted.

Good news and bad news

The good news is that no northern lion populations have recently been extirpated. Phil Henschel did surveys and found that many of the populations mapped in 2001 were, in fact, already gone at that time, but that those he confirmed still exist today.[6] The bad news is that they are all just holding on, none of them really secure. The lion numbers in Senegal – where Phil now runs the Panthera project in Niokolo-Koba National Park – are increasing but are still below 100 lions.

The only West African lion population that was believed to be relatively safe a decade ago occupied the W-Arly-Pendjari or WAP Complex on the border between Benin, Niger and Burkina Faso. This 25,000km^2 area then was home to just over 400 lions. It is not so today.

The area had been supported by international organisations and the infrastructure and policies were relatively well developed. For decades this was the best safari destination in the region, with lots of wildlife. In 2013, I supported the plans of the authorities in Benin to set up a Lion Guards project, with funding from the National Geographic Society. This was before the introduction of SMART, the application that many parks use for monitoring nowadays, so at the time this was adequate.

The Lion Guards had a motorbike and would go round the park, warning livestock herders and keeping tabs on the lions. We would plot their GPS tracks and, together with park authorities, make sure patrols worked for lions. They could call in anti-poaching patrols if they ran into serious issues; and when they found livestock inside the park, they simply engaged with the herders and persuaded them to avoid the park.

Then, in 2019, tourists were kidnapped in a terrorist act. They were later rescued but their driver was killed. In neighbouring Burkina Faso, many outposts and police stations were attacked and law enforcement effectively stopped. The entire WAP complex was listed as 'red', and subject to a high risk of attacks by violent extremist organisations, which meant no travel for tourists.

By this time, management in the Benin sector of the complex had been delegated to African Parks, and the Lion Guards continued operating with them. However, now they had to be accompanied by a soldier on a second motorbike. Towards the end of 2021 we collared lions in the same area where, two months later, a patrol vehicle ran into a land mine, killing eight occupants and wounding 12 staff members. This was just one of many incidents that took place. Today, the situation has deteriorated to the point where law enforcement in the park has become an issue for the military. On the Burkina Faso side, local farmers have been evicted from their villages by the military, and everyone in the area who is not military is considered a suspect.[7]

Whatever the backgrounds of the particular organisations involved in these attacks, they fit into a pattern of terrorism, insurgency and instability rising across the Sahel. Wildlife is too scarce to be a significant source of funding for terrorism, but indications are that some of it ends up in the pot anyway. The WAP was never a place of 'green militarisation', but reality caught up with the Lion Guards. Under current conditions it is difficult to conduct complete and consistent monitoring so we cannot know the situation for sure. But, without extraordinary efforts, the remaining WAP lion population is unlikely to survive the next decade.

Human-wildlife conflict

There is a lot of armed conflict, but there is also considerable human-wildlife conflict. This has always been a key issue in lion conservation. Lions kill cattle and then people kill lions. In 1992, after I had done my master's on human-elephant conflict in Waza National Park, Cameroon, I wanted to carry on and do my PhD there. The obvious priority was human-lion conflict, and this is how I got into lions, opportunistically and to address a problem, rather than as a childhood dream.

I had been a Dutch city boy and the learning curve was steep: mastering French, acquiring bushcraft, learning to communicate with local communities and training my immune system the hard way. I loved the new life and stayed in Cameroon until 2001, with support from Leiden

Bouba Ndjida National Park, Cameroon. © James Kydd

University and my mentor, Hans de Iongh. During those years I experienced human-lion conflict up close.

We had collared five lions, with old-school VHF (very high frequency) transmitters: no sitting in an office downloading satellite data but going out and locating the animals. We learnt about their largely solitary lifestyles, their nocturnal behaviour, their seasonal patterns and their appetite for livestock.[8] We would track them all day and, when we found them on a cow carcass, we would record this and go to sleep in the nearest settlement, probably under the roof of the hapless cow's owners. People were always open to discussing the issue, never hostile towards 'you and your lions'. We tried introducing various mitigation measures, but adoption was low. Then again, so was retaliatory killing.[9]

Over time, I understood that livestock depredation was one of so many challenges to pastoralist livelihoods. Amid cattle raids, agricultural encroachment, human and livestock disease and lack of water and schools, lions were just another nuisance, only partly balanced by the awe they inspired. Some small recompense was that the life of a herder can be dull and the occasional visit by a lion research team offered great entertainment.

Their Muslim faith, in combination with traditional beliefs, taught them that the best way to change destiny was to engage in prayer and ritual, rather than changing husbandry practices. Only when things got out of hand would people use their homemade guns – muzzleloaders that shot ball bearings. The point of the discussion was not about the effectiveness of rituals but about the need to address problems more broadly than just dumping a boma (enclosure) or kraal on local communities to protect livestock at night.

One of the collared lions eventually got shot in the front leg and we patched him up. Checking on him the next day involved crawling on all fours through the same thick acacia bush we had seen him enter, probably the stupidest thing I have ever done. When we came out the other side with nothing but thorn scratches, it turned out my local guide thought it had been madness but trusted me to know what I was doing, exactly what I had thought of him. Though potentially disastrous in this case, trust is a good thing and it had given me valuable insight into local culture. It was a culture that values lions as part of a healthy ecosystem and tolerates depredation, but only up to a certain point.

With rising numbers of people and livestock on ever smaller pastures, pastoralists will increasingly encroach on protected areas and the problems will escalate. However, a solution must primarily address the causes and not just the symptoms. It must also be region specific as the challenges in West and Central Africa are completely different from those in East Africa, where the pastoralism and lion conflict is more constrained in terms of the extent of ranging patterns and complexity of the actors involved.

Nowadays, most herds across Central Africa are not owned by local pastoralists, but by foreigners running from violent extremism and ethnic persecution in West Africa. Wealthy urban elites are involved, too, entrusting their cattle to local herders.[10] When rangers confiscate cattle in the parks, wealthy owners prevail upon them to look the other way.

Lack of trust in modern financial institutions and the prestige of owning cattle are two factors that have boosted the trend towards absentee ownership of livestock. With no one knowing exactly who owns what, opportunities exist for money laundering and the invisible accumulation of wealth. This extensive communal grazing system with high mobility across thousands of kilometres is giving way to intensification and this will affect the protection of livestock against carnivores.

All projects to improve animal husbandry practices, create bomas, introduce lion lights and other deterrents, educate children, promote community resilience and so on are noble but presently insufficient. They must be accompanied by measures that promote a transition in the livestock sector, which makes more money for more people with fewer cattle and less encroachment on protected areas. There will always be residual conflict between lions and locally owned livestock, but there will also always be some degree of tolerance.

Always a glass half full

Many large donors and non-governmental organisations (NGOs) turn away from the northern lion, considering its conservation a highly risky investment, if not entirely a lost cause. It is a shame because our goal in the area is the conservation of lions in functioning ecosystems across their range, not just in fenced reserves in popular, more stable countries.

Africa is changing rapidly and the 'out of Africa' vision of endless open wild plains does not exist, and probably never has. In the near future, Africa's population will double, be largely urbanised and have a consumption pattern that includes holidays to protected areas. No African country can afford to conserve all wildlife everywhere, but most can afford to have some 'conservation jewels'. Good examples are Sudan, South Sudan and Ethiopia, although all have their own history of political turmoil, and are largely neglected by conservation NGOs. This is the extreme eastern end of the range of the northern lion, and the area contains a mix of the two subspecies, each unique.

For many years it was thought that the Serengeti-Mara migration of two million wildebeest, zebra and other species was the largest wildlife migration in Africa. Because it was so seldom visited, it was only recently discovered that in fact the largest migration – around 10 million strong – was by white-eared kob and tiang moving between Boma and Badingilo in South Sudan and Gambela in Ethiopia.

Wildlife collars are standard kit to monitor wildlife movements, and Tanzania and Kenya are full of them, operated by hundreds of research teams that fit them routinely. When we wanted to collar a lion in Gambela a decade ago, Ethiopian authorities were so unfamiliar with this technology that we needed clearance from the security services to import them, and a radio licence for the frequencies of the collars. To my knowledge, 'our' male remains the only lion ever collared in the system, and still no one has a clue how many lions inhabit the region. Recently, African Parks has been delegated to manage this landscape and things will soon change.

On the border between (North) Sudan and Ethiopia is another area that is seldom visited and has hardly been studied: Dinder NP in Sudan and the adjacent Alitash NP in Ethiopia.[11] Despite all the international meetings and networking, this area had not been flagged as a lion area until I visited in 2016 and caught a lion on a camera trap. For a day it was all over the international media. You would think it would have attracted some research projects but it didn't, and one presumes the lions are still there.

The most striking example of international neglect comes from Borana National Park in Ethiopia, on the Kenyan border. There are many endemic bird species there that attract a handful of ornithologists, and some Ethiopians visit the soda lake, a spectacular caldera. Other than that, there is no tourism. It is one of just three places you can see Grevy's zebra, a unique species with narrow stripes and a white belly. The park is also listed as having a complete large carnivore guild, but there is not much supporting evidence.

I went there with the first rains in 2023 after a three-year drought. Dead livestock was everywhere; 1.5 million cattle had not survived the lack of food and water. But the zebra were still there. I could not get a sense of how many but they were not particularly scarce or skittish – clearly, no one had ever thought of poaching a zebra.

People had been allowed to graze their livestock in the park and were happy to do so, despite the occasional goat being taken by a cheetah. Even without financial benefits from wildlife, the community was happy to live alongside it. Lions, however, were a bit too much for the pastoralists and had been chased to the most remote corner of the park, where they remained. Will they be able to regain ground after the end of the drought and recovery of the park? There will be no one to document it but I bet they will. In this region, nothing makes conventional sense – commercially or politically – but the jewels are there for those who want to see them.[12]

In the end, while the status of the northern subspecies as a whole is much worse and deserves far more attention, individually northern lions are perhaps no more threatened than southern lions. Those that remain have done so against the odds and the fittest have survived. Many populations of southern lions still have to weather the storm that has already swept across West, Central and the Horn of Africa and, proportionally, they may still suffer the greatest losses.

'Overall, I am convinced lion numbers will continue to decline for some time, but I remain hopeful that jewels will remain scattered across the continent.'

Hans Bauer

Zakouma National Park, Chad. © Steve Turner, Origins

CHAD

07

Building a lion stronghold

The lions of Chad are in a precarious balance between a stable birth rate in protected areas and loss to conflict with humans outside of these areas. In Zakouma National Park they are thriving.

Chiara Fraticelli

Zakouma National Park is one of Africa's brightest success stories. Created in 1963 and – since 2010 – managed by African Parks in partnership with the Government of Chad, it's a 3,054km² open-boundary park in the Sudano-Sahelian ecosystem of southern Chad.

Because of challenging environmental conditions, for much of Zakouma's history, the active protection and management of wildlife was limited to the eight months of the dry season, with minimal engagement during the four-month rainy season.

After years of intense illegal wildlife killing, by the turn of the century the park's situation was dire. But after increased efforts for its protection by African Parks, in partnership with the Chadian Government, Zakouma recently celebrated five years with no elephant being killed and a general population increase across the board for large herbivore species.

The park also has the highest density of lions in Chad, currently numbering around 150 individuals. Its population was surveyed in 2003–2006[1] and then again in 2013[2] and both studies estimated lion numbers at around 140 individuals. On the basis of this data, the lion population appears to be mostly stable, with relatively high lion densities compared to other West and Central African countries. However, considering the surface area of the park and the prey counts from aerial surveys in the dry season, the park's lion population seems lower than it could support.

One possible factor limiting the growth of the lion population when considering the size of the park and the amount of prey is Zakouma's unique geography and seasonal environmental fluctuations. The national park sits in a low basin that partially floods in the wet season and drives the movement of several wildlife species beyond the park boundaries. In view of this, the true available land area and prey density are significantly lower during the wet season than the surveys carried out in the dry season show. Conversely, the dry season is extremely dry and pushes animals to gather around the few pans in the east of the park, while the west has almost no access to water. Considering seasonal availability of prey, water and land, and if geographically limited to the national park area, the lion population of Zakouma is probably at its maximum.

Of course, Zakouma is not alone: 21% of Chad's surface is covered by protected areas. In the south of Chad, the Greater Zakouma Ecosystem encompasses Zakouma National Park (3,054km²), Siniaka Minia National Park (4,643km²), Bahr Salamat Faunal Reserve (20,950km²) and the wildlife corridors employed by the wildlife of these parks. Together, these areas make up around 45,000km² and constitute one of the best starting places for robust lion conservation.

66 ~ THE LAST LIONS

Zakouma National Park, Chad. © Michael Lorentz, Safarious

Camp Nomade, African Parks' seasonally operated tented camp, is erected for just a few months every year. It is located deep within Zakouma National Park, Chad, and overlooks a seasonally flooded plain that attracts a wide variety of wildlife to their front 'door'. © Michael Lorentz

Chad is also a land of movement, with nomadic herders walking between 100km and 400km north to south and back every year with their cows, sheep, goats and, more recently, camels. Wildlife, driven by seasonal floods and droughts, must migrate a long way between their dry and wet season ranges. GPS collar data shows that most hartebeest and tiang migrate more than 180km north out of Zakouma every wet season, looking for drier areas.

Historical records show that this might be an underestimation of their traditional migration distance. The elephants from Zakouma used to migrate more than 130km north towards Abou Telfan and Heban, and west towards Siniaka Minia during the wet season. While nowadays their movement is more limited, they still migrate a good distance out of the park's boundaries. Their migration paths pass through areas of varying protection status, under varying management and often into vulnerable areas near human settlements, to the detriment of both humans and wildlife.

Lions do not migrate, but shift seasonally within the territory. During the dry season, they locate themselves on the floodplains and waterholes inside Zakouma. In the wet season, they are forced to move in search of prey that decreases in density when the antelopes migrate. This territory shift may also be linked to the avoidance of people. During the dry season, nomadic herders and fishermen are widespread across the Greater Zakouma Ecosystem, leaving only the national park as a safe refuge. Competition at waterholes outside of the park is high, and lions, like other wildlife, tend to prefer the territory inside of national park boundaries.

With the wet season, the nomadic herders move north but so do the wild prey, at the same time forcing the lions to travel further in search of food. This increases the risk of interactions as the low density of wild prey pushes lions to take the livestock of permanent villages located around the protected areas. Although reports of human attacks by lions are extremely rare, depredation of livestock is common and – as in other parts of Africa – is linked to killing of lions, either as prevention or retaliation.

The bigger picture is more complex because not all lions are Zakouma residents; and without the option of a safe refuge, they are more vulnerable to conflict all year round. It's not uncommon to hear of lion sightings outside of the park, even along busy main roads. Some reports are from our community teams working in the area, others are from those who have lost livestock to lion attacks.

Escalating global and local disruptions – COVID-19, insecurity around the sudden death of the long-term president of Chad, Idriss Déby, increasing pressure on natural resources and climate change – are forcing up living costs and reducing socio-economic security. These factors are contributing to the decreased tolerance towards carnivore attacks on livestock and a general increase in illegal killing outside of the main protected areas.

As lions are killed, especially in areas that are vulnerable to seasonal conflict between humans and wildlife, space opens for new lions to move in. These will be young ones dispersing from Zakouma, which leave the safe refuge to look for a new territory and a pride. This brings them into contact with the same risks and conflicts the previous resident lions had, raising the possibility of a vicious cycle limiting lion population growth in southern Chad.

For this reason, community engagement, environmental education and conservation-oriented actions are key to mitigating conflict. In the Greater Zakouma Ecosystem, African Parks has a growing community department focused on this work. Prevention and mitigation techniques to reduce human-carnivore conflict are being trialled to see what is best suited to individual communities.

Today, Zakouma is a safe space for lions and other wildlife, while neighbouring community lands are prone to conflict. With Siniaka Minia having been raised to National Park status in early 2024, the work is underway to develop this area as a second core protected area for lions and other wildlife. Sustainable and conservation-based land use plans for the Greater Zakouma Ecosystem can be developed to promote coexistence and decrease the loss of lions beyond park borders.

With two well-managed protected core areas and effective conservation-oriented management and coexistence in the surrounding community areas, Chad has the potential to become a lion stronghold in the future. Our goal is to work towards conservation and coexistence, in the hope that our cubs will grow into strong, healthy lions and be able to live and thrive in the wider ecosystem of southern Chad.

The plains of Zakouma National Park, Chad, are renowned for their extraordinarily rich concentrations of wildlife. © Brad Hansen

CAMEROON

08

Bouba Ndjida National Park

Up against cattle farmers and hunters, the lions of northern Cameroon are hanging on, supported by the efforts of a few dedicated conservationists.

Dr Paul Funston

In West Africa, national parks and protected-area networks are generally small, seldom larger than a few thousand square kilometres. But in Central Africa, by comparison, there are some vast conservation areas such as the Zakouma in Chad, the Chinko wilderness in the Central African Republic and the huge conservation estate of South Sudan where massive migratory herds still roam. Until quite recently, the Bénoué Complex in northern Cameroon held the best numbers of lions in Central Africa. Biologists suggested that the complex might have supported as many as 500 individuals – truly remarkable by West and Central African standards, and a prime reason why the Bénoué Complex deserves vastly more attention than it gets.

These lions lived in a vast protected-area complex reaching across borders, from northern Cameroon and Nigeria in the west to Chad in the east. At about 30,000km² or three million hectares, it's a very big area. But here's the thing: as with any protected area in much of Africa north of the equator, conservation investment is poor, safari hunting is held up as an economic cure-all and conservationists and their organisations live in blissful and naive hope.

Although the Bénoué Complex is large, it is comprised predominantly of hunting zones, or *zones de chasse*. In the complex lie three national parks, each of which is about 2,000km² – Faro in the west, Bénoué in the centre and Bouba Ndjida in the east. National parks are supposed to be the source areas inside any protected-area complex, safe zones buffered by hunting zones where wildlife occurs at lower densities.

Sadly, as with so many national parks in Africa, there is in reality little distinction between national parks and hunting blocks. In Cameroon all are depleted, heavily poached, grazed by livestock and subjected to illegal logging and mining.

Having radio-collared several lions in Bénoué National Park 20 years ago, I was dismayed on a recent trip to learn that the current conservator appears corrupt and condones all of the above

assaults on Bénoué National Park, which is in steep decline. And the consequences are disastrous: prey is depleted to such an extent that no lions can realistically survive there.

This story is all too familiar to those of us involved in surveying or radio-collaring lions in the disaster-zone parks of West Africa. Two years ago, I was dismayed to have to declare lions extinct in Yankari Game Reserve – and most likely in all of Nigeria. Dismal prospects for lions, and thus for people too, is what I have come to expect in the region.

However, it was with some optimism that I recently made the first-ever attempt to radio-collar lions in Bouba Ndjida National Park in Cameroon. And what a surprise! The park had almost boundless numbers of general game, probably the biggest population of giant (or Lord Derby's) eland in the world, and it offered the real prospect of prides of lions.

And we found lions in Bouba Ndjida: a solid 40 individuals, and possibly as many as 60. At over 40 degrees Celsius every day, it was not easy, the heat debilitating. Worst was at night, sitting on the back of a pickup staring into the darkness with my night-vision scope, almost not daring to breathe. That is how difficult it was to dart these lions, and, after 10 days I had still caught nothing. However, a younger, fitter and altogether incredibly dedicated vet, Dr Richard Harvey, had caught and collared four. We soldiered on and, once 'the competition' had moved on to dart forest elephants somewhere else in Cameroon, we found our form. There are now 11 lions in Bouba Ndjida wearing radio-collars, which are hopefully contributing to their survival.

Why is Bouba Ndjida succeeding in the face of such dismal conservation performances almost everywhere else in the region? Why is there prey, lots of prey, and even some surviving lions? The government and NGOs are pulling together, but there is also one incredibly committed man: Paul Boor. He has dedicated his life to this one park, working tirelessly on his own for years; and he has brought in the Wildlife Conservation Society (WCS) to co-manage the park, together with a dedicated and supportive conservator, Serge Patrick Tadjo. As a team, they have the determined goal of restoring the park and ultimately rehabilitating the conservation landscape.

The challenge is immense. Fulani cattle herders are feeling the squeeze of climate change and the instability of violent incursions by Muslim fundamentalist groups. The park managers are doing an incredible job, sometimes literally having to dodge bullets. Lions are still killed in numbers outside or towards the edges of the park, but there is a core population of about 40 surviving the odds by breeding as fast as they can. There are lots of cubs, few subadults (possibly the most vulnerable subset), a core group of wiley, clever lionesses and some spectacular male lions able to outsmart their attackers by being shy and reclusive.

Lions at the heart of the park, however, are quite tame and these we radio-collared. They are the focus now of a study, and their movement data can be used to design more effective anti-poaching patrols and strategies. They need to be closely monitored by a team ready to adapt its efforts in response to seasonal movements. Solving the issues the Fulani have with lions will be a far greater challenge, but one that must be tackled if any form of restoration is to be secured.

Bouba Ndjida and Bénoué are linked by hunting blocks, which also extend to Faro – a failed construct in my view. The hunting blocks are so depleted by poaching and cattle grazing that enormous efforts will be needed before a core lion population in Bouba Ndjida can recover across a larger landscape.

In fact, this recovery may never happen, which is my current prognosis. But individuals such as Paul Boor and Serge Patrick Tadjo, supported by WCS and global funding agencies, may yet prove me wrong. How dearly I hope so, and that lions will once again roam and roar throughout much larger areas in Cameroon. If we can succeed there, then surely it's possible to do so in many other lion landscapes across the 'hopeless' protected areas north of the Equator.

© Dr Paul Funston

ETHIOPIA

09

The elusive lions of Ethiopia

Lions were a symbol of royalty and the memory lingers on, but they have been pushed to remote areas and need urgent protection.

Graeme Lemon

The hectic, familiar clamour of Addis Ababa rippled around me as I sat with a friend and fellow guide, catching up in a downtown coffee shop. The capital's hectic energy barely registered, though. Our conversation had turned to lions and I was absorbed as my friend described his experiences in the remote far west of the country.

For several years he had lived in this region and, on occasion, had ventured out on day trips to trek through the bamboo forests south of the town of Assosa. To his amazement, he had sighted lions moving through the forest. After that first sighting, he and his friends went on regular excursions to the area to marvel at these unexpected residents. I wondered if these were the 'lost' lions highlighted by the Born Free Foundation in 2016. Then I remembered those lions were photographed in Alitash National Park, 200km north of Assosa.

Finding lions in the bamboo forests was an exciting discovery, particularly since they were in a region well beyond any protected areas. Wild and often underexplored regions still exist in Ethiopia, so it's not that surprising that big cats still roam in remote parts.

Symbols of royalty

This ancient land has always had a special relationship with lions. Gondar – famous for its 16th century palaces – offers clear evidence of lions being an element in royal history, possibly as symbols of regal power and strength. The palaces built during the time of King Fasilides feature well-built lion cages and it has long been suggested that the kings kept lions as pets. Even in the late 1800s, Emperor Menelik the Second was rumoured to have taken lions into battle with him to help repel the invading Italians. Interestingly, according to folklore, bees were also employed as part of his battle strategy.

The last Emperor, Haile Selassie (known as the King of Kings), kept lions on the palace grounds in Addis Ababa. They were prominently displayed and paraded to the public at every opportunity. On one occasion, Selassie's exotic pets were on display at the grand opening of the Hilton Hotel in Addis Ababa and were taken to the ancient town of Lalibela for a photo shoot to mark the occasion. His lions, genetically distinct from all other lions, and with impressive dark manes, even made an appearance in a Hollywood film.

Of course, much has changed since the death of Selassie and Addis has undergone considerable development. But as recently as 12 years ago, I was told that the late Emperor's lions could be found on the university grounds. Highly improbable, but this belief does illustrate the enduring memory of the royal lions. There remains a special place in the heart of this country for its big cats, so often echoed in emblems and ceremonial devices.

Emperor Haile Selassie kept lions at his palace.

Addis Ababa. © Ariadne van Zandbergen

Prides of the plains

It is not hard to imagine that lions were once a common sight in forests and on the savanna in this vast landscape. Awash National Park, one of the first protected areas in Ethiopia, is three hours' drive to the east of the capital and was once famous for sightings of lion prides on the open plains. Further to the south, the Bale Mountains National Park was known as the home of unusual black-maned lions. In fact, the country is renowned for this particular subspecies and these majestic cats still prowl the dense Harenna Forest, occasionally sighted on the single track that runs through it.

Yet, the development of modern Ethiopia has certainly taken its toll. According to estimates, between the 1980s and now, the human population in Ethiopia has trebled and it is one of the fastest-growing populations in Africa. The constant need for more land and food crops means former wilderness areas are being overtaken and turned into subsistence farms. Protected areas are now under threat as pastoralists push into remote habitats, forcing humans and wildlife into conflict.

Successive Ethiopian governments have faced this challenge by creating a diverse array of protected conservation areas. Today, there are over 30 such areas and national parks, which account for 14% of the country. Despite this, law enforcement in these locations is fraught with difficulty and once-beautiful parks such as Awash have become the domain of pastoralists.

Remote lions

For the 12 years I have lived and travelled in Ethiopia, conversations with wildlife officials and conservationists indicate that there are viable lion populations living in remote regions. The greater Omo Valley and the area between Mago and Omo national parks are still home to prides. On my first 10 trips into the Omo, we sighted lions roughly half the time.

One of these sightings was an unforgettable evening walk with a local tribesman, when we found ourselves 20m from a snarling lioness as she emerged from a dense thicket. I noted that my companion, though unarmed, did not flinch in the slightest. He merely turned and grinned from ear to ear as the lioness bounded away from us. Sadly, the tribesman's complete lack of fear was a warning of the changing relationship between lions and local inhabitants. Because villagers had pushed the lions out, the cats were no longer seen as the deadly threat they once were.

A few years ago, a trip to see the kob migration in the wetlands of Gambella offered daily sightings of lions. Given the vastness of the terrain and the different locations of the lion sightings, I estimated there must have been some three or four different prides. Gambella is a wetland wilderness; so, for a good portion of the year, it is not an ideal habitat for lions. In the circumstances, these prides undoubtedly lead a nomadic existence, following the kob herds between Ethiopia and the savanna plains of South Sudan.

Around 30–40 years ago, lions were found on the northern grasslands and along all the boundaries of the Bale Mountains National Park. Currently, they are restricted to the deep, untouched habitat within the Harenna Forest because of an increasing population putting pressure on the park's boundary. Unfortunately, the forest is no longer a guaranteed safe haven. Pastoralists and honey collectors are becoming an ever-present threat as they push deeper and deeper into this once-pristine wilderness.

Lions found in the Harenna Forest have the same distinctive black manes as those once found in the emperor's palace grounds. DNA sequencing shows they are genetically unique. Given that two other subspecies have already become extinct in Africa, there should be a significant focus on protecting the remaining population here.

On a trip to the remote Wabe Shebelle gorges 10 years ago, I found a dry and unpopulated landscape broken up by three deep river canyons. Separated by just 20 or 30km, these gorges stretch east towards the Somali province. Greater kudu and Hamadryas baboons offered fantastic sightings in the late afternoon and we were serenaded long into the night by several lions calling quite close to our camp. At that time, all three gorges were completely unpopulated by humans, though we did find tracks of a pastoralist and his cattle in one of the gorges.

Some years later we were exploring further south in the same area by helicopter, and we put down at one of the permanent pools along the riverbed. Barely out of the chopper, we came across the large pug prints of a big male lion, his tracks imprinted in the soft, sandy soil indicating he had recently passed by. I was amazed to find lions still roaming freely in this wild area with no designation or protection. A recent journey through the area by road, however, told a different story. It was evident one of the river gorges had been taken over by commercial farmers and I noticed a tractor ploughing close to where we had originally camped and listened to the lions calling into the night.

Wildlife destination

Ethiopia will never be a destination for wildlife watchers who expect to see multitudes of wild animals crossing the dusty plains. It's a specialist destination, home to the kind of endemic species that would enthral only a true naturalist. There are impressive plains, though, and these offer opportunities to see several different mammal species. One of these is the Halledeghe Wildlife Reserve with its large landscape of open grassland on the southern extreme of the Danakil Desert. Lion and cheetah sightings used to be regular there and lions are still occasionally sighted, but there are now no cheetahs left, mostly a result of traffic deaths when they tried to cross the nearby Addis Ababa to Djibouti highway.

Regrettably, the Halledeghe wilderness area has a hunting block at its northern boundary and the Djibouti highway separates the two areas. Within this concession are permanent water sources, and wildlife from the plains gravitate to the concession during the dry months. In a country fast relinquishing its large cats to massive habitat loss, it is disturbing that lions are still included in hunting quotas here. While there needs to be an understanding for the actions of pastoralists who kill lions to protect their livestock, the potential demise of one of Africa's most iconic animals through commercial hunting is highly controversial.

Fortunately, the land beyond the hunting concession is increasingly uninhabited to the north and east, eventually becoming the hot, dry Danakil Desert. This inhospitable space offers hope and a possible refuge for the remaining lion population in the area.

Another wild area is between the eastern Oromia and the western Somali regional states; a vast dry landscape intruded on only by a pair of snaking rivers. According to Google Earth, this is an untouched space of around 5,000–6,000km^2, which will undoubtedly be home to a number of big cats. The turbulent relationships between the ethnic and tribal groups that border this landscape have effectively protected it and the area continues to remain relatively free of human habitation. For this reason, it is still a sanctuary where wildlife can exist and survive, though pastoralism is creeping in.

I used to think lions and pastoralism could not mutually exist but have witnessed evidence to the contrary. Laikipia in Kenya appears to have found a balance, as the area is home to large tracts of land with both cattle herds and lion prides. The cattle are well protected and housed in electrified bomas at night. The success of this practice suggests that creating communal bomas may help ensure the survival of Ethiopia's lion populations. But when an apex predator is involved, the boma will not easily remedy the difficulties created when agricultural livestock share their valuable wilderness lands.

A decade ago, it would be rare to see a tractor as all tillage was done by hand and oxen. Today, tractors have sped up the development of previously untouched lands. The combination of commercial farming for local food production and the desire to generate foreign currency through exports has rapidly turned the once-pristine Omo and Gambella landscapes into highly threatened habitats.

This begs the question: in a country that needs to prioritise human requirements, what happens to the big cats that once roamed the plains? The future for this symbol of African pride in Ethiopia is tenuous and the few remaining prides in the national parks and protected wildernesses are certainly facing extinction. However, hope remains.

When Ethiopia achieves political and economic stability, as it must, protected area management will improve. Even now, past and present leaders of the country are focused on better management of wilderness resources. Exposure and publicity will help reinforce the urgent need for protection. Support from conservation bodies will also ensure that local wildlife authorities have stronger influence and control over protected areas, helping them stem the tide of habitat loss. The time to act to protect these prides has almost passed and the urgency of this crisis is extreme. Before the roar of the Ethiopian lion is permanently silenced by the grinding gears of agricultural machinery, we need to find a way to save the symbol and pride of Ethiopia.

Gambella National Park. © Steve Turner

CENTRAL AFRICAN REPUBLIC

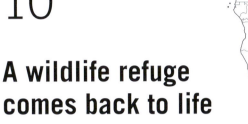

10

A wildlife refuge comes back to life

Deep in the Central African Republic (CAR) lies a little known wildlife refuge that is, once again, starting to thrive. In the far eastern region – one of Africa's most remote and volatile parts of the country – Chinko Nature Reserve is a remarkable story of hope and success.

African Parks

Lions have started to make a comeback in Chinko Nature Reserve, Central African Republic. © African Parks

Since 2014, when the CAR government invited African Parks to manage the refuge, wildlife has begun to return and stability is becoming a reality in a region that was once fraught with civil war, poaching and lawlessness.

The Chinko team's efforts to protect this landscape, to work with communities and to deflect herders from illegally driving cattle through the conservation area, has led to significant reductions of poaching, and has created one of the most stable places in the region.

Because of these successes, in 2020, African Parks signed a revised agreement with the CAR government for another 25 years, which has increased the area under management, including the Functional Landscape of Chinko, from 6,000km^2 to over 64,300km^2.

Today, Chinko has a growing population of elephants; 23 confirmed carnivore species, including endangered African wild dogs; northern lions; nine species of mongoose; 14 primate species, including eastern chimpanzees; 24 different ungulates; three pangolin species; nearly 500 bird species and over 100 species of fish.

A wildlife refuge comes back to life ~ **81**

SOUTH SUDAN

11

A newly rediscovered migration

In mid-2024, African Parks announced the results of a year-long aerial survey undertaken in one of their most recent management projects, South Sudan's Boma and Badingilo national parks.

Marcus Westberg & African Parks

It's difficult to know what was more astonishing – the census's findings that some six million white-eared kob, tiang, Mongalla gazelle and bohor reedbuck migrate across the Boma Badingilo Jonglei Landscape every year, making it by some margin the biggest migration of large land mammal migration on Earth, and one of the longest – or the fact that, until now, we had little idea it was there and no idea of its magnitude.

Comprising nearly three million hectares of mostly unspoiled wildlife habitat, Boma and Badingilo are an integral part of a vast, 20-million-hectare wilderness that extends to the east and north of the young country's capital, Juba. Decades of political instability and civil war have meant little to no active conservation. But this also meant almost no infrastructure development and a low, subsistence-based human population outside of the major towns and cities.

Over the years, while the migratory antelopes have escaped sustained persecution, sedentary species have been hunted for meat if not for profit. Nevertheless, most of them remain, albeit in smaller numbers than this land can carry. Populations of zebra, elephant, buffalo, eland and Nubian giraffe, as well as those of carnivores, are poised to recover and thrive now that African Parks has begun its watch.

While carnivore numbers are unknown, spotted hyena, leopards, cheetahs and wild dogs are present in the area and lions have been regularly seen in around 60 aerial observations since African Parks began its operation in 2022. Four individuals have been collared, the hope being that this will significantly increase the understanding of the movement and behaviour of the lions there.

It remains to be seen what the future holds in store for the Boma and Badingilo national parks. But with African Parks' proven track record and approach, local communities eager for development and the South Sudan government's determination to turn the country into a wildlife tourism flagship – not to mention the news of the migration, which has become a point of national pride – there is real reason for hope. And if six million migrating antelope somehow managed to escape our collective notice for so long, perhaps there are other overlooked wildlife havens and extraordinary feats of survival elsewhere in Africa. It's a big continent.

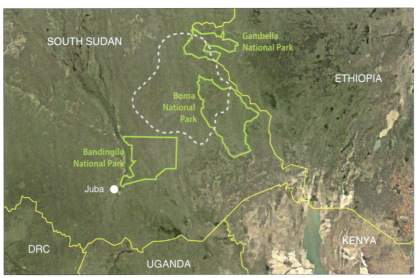

Above: A small part of the newly discovered migration in South Sudan, which is the biggest migration of large land mammals on Earth.

Left: The 825km migration route of an adult male white-eared kob (*Kobus kob leucotis*) that was radio-collared and tracked in 2013 in the Boma-Badingilo-Gambella transboundary ecosystem[1].

A newly rediscovered migration ~ **83**

African Parks conducting their aerial survey of the newly discovered migration in South Sudan.
© Marcus Westberg

A lion is fitted with a tracking collar, which will help scientists learn more about their movements and migration patterns in South Sudan.

The migration is not only about kob; the Boma Badingilo Jonglei Landscape has a richness of wildlife that has gone unnoticed for too long.

Maasai Mara National Reserve, Kenya. © Jeffrey Wu

The lions of Kenya

PORTFOLIO

Maasai Mara National Reserve, Kenya. © Grant Atkinson

Richard Turere, then 13, attaches his "Lion Lights" invention to the timber posts of a cattle kraal close to his home on the outskirts of Nairobi National Park. Richard devised these lights when he was 11 years old, and they are an effective local solution to lion attacks on Maasai livestock at night. "These lights are a small device which protects Maasai cows from lions," says Richard. "Lions fear moving lights, so I have made a device that tricks lions into thinking that I am awake and walking around when I am actually sleeping in my bed." Richard has become a minor celebrity and has given a number of TED talks about his invention. © Brent Stirton/Reportage for National Geographic Magazine.

Maasai Mara National Reserve, Kenya. © Michael Poliza

'Throughout my career I have tried to capture these breathtakingly beautiful, heart-stopping, thrilling moments in nature and, more importantly, I have tried to use the power of these compelling images to bring an awareness of conservation to my viewers.'
───────────
Jeffrey Wu

Maasai Mara National Reserve, Kenya. © Jeffrey Wu

PORTFOLIO

Jeffrey Wu

Jeffrey Wu is a Canadian professional wildlife and nature photographer accredited by Professional Photographers of Canada (PPOC). He is an artist, author, educator and a conservationist.

Jeffrey Wu was born and raised in Shanghai, China. His love of photography was nurtured by his mother, who was a professional photographer and taught him photography from the age of seven. Jeffrey immigrated to Canada in 1993 at the age of 27. In 2012 he went to Kenya on a photography trip that changed his life forever. In 2013 he sold his three restaurants in Toronto, became a full-time professional wildlife photographer and spent the next 10 years building his career, photographing wildlife in the Maasai Mara National Reserve, Kenya.

Jeffrey Wu is also a world-renowned photographic competition judge. He has judged some of the most prestigious competitions of the world, including the 2018–2019 Nikon Photo Contest and the 2017–2019 Nature's Best Photography Africa. Jeffrey's photographic passion is African wildlife, leading him, since 2014, to spend 10 months of the year mainly in Kenya and other African countries.

In 2021, Jeffrey became Kenya Tourism Board Brand Partner.

Maasai Mara National Reserve, Kenya. © Jeffrey Wu

Maasai Mara National Reserve, Kenya. © Jeffrey Wu

Maasai Mara National Reserve, Kenya. © Jeffrey Wu

Mara River, Maasai Mara National Reserve, Kenya. © Jeffrey Wu

Maasai Mara National Reserve, Kenya. © Jeffrey Wu

Olkirimatian Conservancy, East Rift Valley, Kenya. © Jeffrey Wu

Maasai Mara, Kenya. © Angela Scott

12

Maasai Mara – a fragile Eden

Following a single pride for more than 45 years in the Maasai Mara National Reserve has given us a privileged insight into the state of lions in East Africa.

Jonathan & Angela Scott

We have followed the fates and fortunes of the Marsh Pride since 1977. Our base at Governor's Camp is set close to the glorious Musiara Marsh, which gave the Marsh lions their name and is the heart of their dry-season territory. To the east, the intermittent watercourse known as the Bila Shaka lugga has always been the pride's traditional breeding site and resting place. Bila Shaka means 'without fail' in Swahili, testimony to the fact that you could always find lions here. This is no longer the case.

A recent 90-minute television documentary, produced by BBC/PBS, called *Lion: The Rise and Fall of the Marsh Pride* revives memories of an event that shocked the world, drawing on a potent mix of voices to narrate the story and give it balance. An earlier award-winning television series, *Big Cat Diary* (1996–2008), had garnered legions of fans and made household names of the lions involved.

On Sunday, 6 December 2015, news broke of the poisoning of eight members of the Marsh Pride that had killed cattle encroaching illegally inside the reserve at night. A decoy carcass had been laced with a highly toxic carbamate pesticide that is still freely available in Kenya. Three of the lions died: Sienna (aged 10), Bibi (aged 17) and a young male named Alan.

The Marsh Pride had always been particularly vulnerable to conflict as its territory spreads beyond the reserve boundary. When Musiara Marsh becomes waterlogged, the lions move to higher ground north and east where they can find prey, either wild herbivores or livestock. Losing lions to poisoning or spearing is part of life for the pride and many of the stars of the series *Big Cat Diary* have been speared or poisoned over the years, including the pride male Scruffy and the lionesses Lispy, White Eye and Red.

In 2004, the pride numbered 29 lions, a high point in its history. Owing to pressure from livestock invasions and a shortage of prey, in 2022 the core group of 11 lions abandoned their territory and moved into the Mara North Conservancy. They returned in 2023 but who knows what the future holds?

Maasai Mara National Reserve, Kenya. © Angela Scott

'We need to re-engage with wilderness and to value it as the source of life, as the provider of our fresh water, our food and the air we breathe, and use it to remind people that the world will be a poorer place without other forms of life to share it with and marvel at.'

Jonathan Scott

Lion research in the Maasai Mara

When Professor Joseph Ogutu surveyed the lion population between 1990 and 1992, he recorded 22 prides within the reserve, 12 of them with 20 or more lions. The Talek Pride was the largest at that time, with a phenomenal 48 lions: four males, 17 females, 10 subadults and 17 cubs. A pride of that size is unheard of today, with smaller groups of lionesses now the norm. This pride was later to become known as the Fig Tree Pride and is suspected to have been wiped out by herders at night in 2021. Only three lionesses from a breakaway group are still seen.

Ogutu found that roughly half the Maasai Mara prides were accompanied by two adult males. Their average tenure was two to three years before being ousted by nomadic males or males from adjacent prides, and just long enough for their cubs to reach subadulthood. All young males are ousted from their natal pride at two to three years of age. They roam widely until around the age of four and, given their lack of hunting experience at this stage, view cattle as easy targets. Unsurprisingly, scientists with the Mara Predator Conservation Programme have evidence that many of these young males are killed in conflict with pastoralists.

One consequence is that, with fewer rivals to take their place, pride males may end up staying longer, and likely attempting to breed with their female offspring once they reach three to four years of age, with obvious repercussions for genetic health. Scientists believe that, to avoid this situation, groups of young lionesses are leaving their natal pride to try to find a territory of their own. This is unusual behaviour, and something they might otherwise do only when there are too many adult females in their pride and not enough resources in the form of food and den sites. This – and the fact that some of these young females are being lost to conflict with pastoralists – would explain why we're seeing smaller groups of adult lionesses at the heart of many Mara prides, particularly in areas where cattle incursions are most prevalent.

Lack of sufficient competition from younger males has no doubt played a part in the success of the larger coalitions of males in the northern Mara, such as Notch and his boys (numbering six males), the four Musketeers that included the legendary Scarface, and the six Warriors (a coalition of six males also known as the Contenders or Six Pack), all of which share history with the Marsh lions.

These powerful coalitions controlled huge areas of 400km^2 or more, mating with lionesses from multiple prides, siring large numbers of cubs and terrorising (and at times killing) other pride males. Though the members of large coalitions sire many cubs, they often fail to invest sufficient time with the mothers to protect their offspring from intruding males. Instead, they move between prides – mating, spawning cubs, and moving on again – or moving back.

In the 1970s and 1980s, the pride males we observed were particularly vigilant when there were small cubs sired by them, staying close to the females as well as patrolling the territory, roaring and scent marking. And this was necessary: the threat posed by rival males was ever present and we would regularly encounter nomadic groups of up to nine members strong, particularly when the wildebeest poured in from the Serengeti; a moveable feast that sustained these wanderers while they sought opportunities to oust resident pride males and establish themselves. Today, nomads are far less apparent.

These days some male lions are living to a ripe old age of up to 13 years (such as Scarface, even 15 years in the case of Morani and Notch 2) thanks to the increasing tendency of the reserve management to intervene and treat male lions injured in fights over territory. Many suffer terrible injuries along the way, yet somehow survive. But generally, fewer than 10% of male lions reach what is regarded as old age.

Nature has its own rules. We should respect and abide by these without fear or favour. The robust tenacity of wild lions has been honed through competition to ensure that only the fittest survive and breed. Interfering with natural processes is likely to disrupt this process: treating a lion injured in a bruising battle with other males might enable it to recover and retain its territory rather than being ousted, but doing so denies or delays other lions – younger, fitter and/or more numerous adversaries – their chance to become

Maasai warriors are renowned for their remarkable jumping dance, the *adumu*, which is used to celebrate significant milestones, such as the rite of passage into adulthood. They also have a long tradition of hunting lions, but this is no longer sustainable.

pride males and breed. As hard as that may be for visitors to witness, we should abide by nature's way.

But as Professor Kay Milton (retired Professor of Social Anthropology at Queen's University Belfast, Northern Ireland) commented: 'It doesn't always help to set up "nature" as a realm or system separate from the human world. This points to a dilemma at the heart of environmental discourse. It is sometimes expedient for conservationists engaged in political debate to treat nature as an independent entity with its own rules. But "nature's way" lost its independence when humans (in many ways and at different times and locations) ceased to depend on its processes and started to manipulate them; intentionally (e.g. pastoralism, agriculture) or inadvertently (e.g. climate change). Conservation is itself an intervention in nature's way. So, while it is useful to point out that helping individual lions to live to old age is not a good thing for the wider community of lions, I don't think it makes sense to set this up as an opposition between human compassion and nature's way. The important questions are: what are the problems, what are the causes of those problems, and how can we resolve them?'[1]

The Maasai

People often ask if we were shocked by the poisoning of the Marsh Pride. No. Traditionally, the Maasai were active during the day, returning to their homes with their livestock before nightfall when predators such as lions, hyenas and leopards are most active. Yet, as communally owned land has been subdivided into individually owned plots of 40–60 hectares, the Maasai have become more sedentary, constructing permanent dwellings and erecting fences.

The number of people living in the Greater Mara ecosystem (the reserve plus the Wildlife Conservancies and other privately owned land covering 6,000km^2) has increased exponentially from just over 19,000 in 1962 to nearly 150,000 in 2019 – an increase of some 670%. This represents an average annual growth rate of 11.5% during the past 60 years, driven partly by a hunger for land and partly by the opening up of opportunities in the lucrative tourism industry.

Some landowners have opted to lease their land to tourism partners for a monthly fee per acre, creating Wildlife Conservancies where cattle grazing is permitted on a rotational basis but where predators still thrive, and tourism numbers are strictly regulated to conform with environmentally sustainable practices. Others cling to a purely herding existence, which is their right. However, as Professor Milton cautions: 'Oh, how I dislike this concept of "right"! It's totally unsustainable in a changing world. Who or what confers rights on people, or animals, or anything? God? Tradition? History? We'll never find solutions if we allow our imaginations to be limited by such entities. In political arguments, rights are immovable objects. They are used as negotiating tools because there is no answer to them. Once they are accepted as valid, the potential for finding solutions is severely limited. And yet, many people have been willing to ignore or relinquish them in order to pursue desired goals, including the greater good. I don't see any way forward otherwise.'

However, the settled areas at the edges of the reserve, bordering the Bila Shaka lugga and the Olare Orok and Talek rivers, are of particular concern. Many people living here tend to view the reserve as a source of additional grazing, their cattle camped along the boundary waiting until nightfall when visitors are safely out of sight before moving in. Yet, this is when the likelihood of conflict with predators such as lions and hyenas is at its greatest. Safari guides and conservationists have been complaining about this situation for years.

Deep tracks leading into the reserve are visible from space, along with piles of cattle dung scattered deep within it. Frequent droughts have exacerbated the situation but incursions are no longer limited to dry periods. Barely a day passes when cattle are not being grazed somewhere inside the reserve. From the herdsman's perspective, when a lion kills a cow it is as if someone has hacked their bank account with little chance of compensation.

Many pastoralists question why they should not be able to bring their livestock into the reserve, given that it's not fenced off from the adjoining privately owned land where the wild herbivores are free to graze year-round. Ensuring a transparent and equitable sharing of reserve revenue with the local community would help to offset the negative aspects of living with wildlife and foster a positive attitude to conservation. Unfortunately, massive theft of reserve revenue has been prevalent for many years, with the community receiving only a fraction of the amount they are entitled to.

Several initiatives such as the Mara Predator Conservation Programme (for which we are ambassadors) are helping to address the root causes of the human–wildlife conflict, encouraging the provision of predator-proof stockades constructed from recycled plastic poles and wire fencing and with metal doorways. The installation of solar-powered flickering lights is another highly effective innovation to help deter predators from trying to break into homesteads at night.

However, none of this helps if livestock is grazed at night and lions are poisoned. The Mara's big cats are the bedrock of Kenya and Narok County's tourism industry. The number of sheep and goats has increased exponentially in the Greater Mara ecosystem: by 269% from some 158,000 animals in 1977 to over 584,000 in 2022; while the number of cattle in the ecosystem dropped slightly by 12.8% in the same period from nearly 223,000 in 1977 to about 194,000 in 2022. But the problem is not just about numbers. Livestock has expanded to densely occupy most of the ecosystem, intensifying competition with wildlife and displacing it.

The loss of Kenya's wildlife

Since his pioneering lion study, Professor Ogutu and his colleagues have documented the fate of Kenya's wildlife both in and outside protected areas. They have found that in the past three decades, the country has lost nearly 70% of its wildlife, spanning a wide cross-section of species, including both predators and herbivores. They point to the long-term declines of many of the charismatic species that attract tourists, such as lions, elephants, giraffes and impala.

Most recently, they have examined the status of East Africa's five wildebeest migrations: Athi–Kaputiei, Mara–Loita, Amboseli, Tarangire–Manyara and Mara–Serengeti. Land subdivision, growth in permanent settlements, expansion in agriculture, roads and fences are increasingly restricting migratory routes and access to traditional grazing and calving grounds in unprotected lands. These processes, coupled with increasing human population pressures and climatic variability, are exerting tremendous pressure on these wildebeest migrations.

Kenya's Loita Plains used to be the breeding grounds and nutritious wet-season pastures for the Mara-Loita migration of wildebeest and zebras. During dry seasons, they would move south to the Maasai Mara and meet up with Tanzanian wildebeest from the Serengeti. Today, the migrations are a shadow of their former glory. The area covers about 7,500 km^2, with large tracts of land leased to commercial wheat farmers. The number of migrating wildebeest has declined by 81% from over 123,000 animals in 1977 to under 20,000 in 2016. Professor Kay Milton expressed her concern to me: 'I don't know much about this, but isn't wheat a particularly thirsty crop and therefore a very bad choice for this part of Africa, especially given the insecurity presented by climate change? Is there no oversight of land use at a national and international level (given that wildlife migrations cross national boundaries)? Of course, I would not expect there to be such oversight, given the nature of national economic/political systems. But I suspect the only solutions to human–wildlife conflict on this immense scale lie in management at ecosystem level, with "big earners", such as agriculture and tourism, being managed for mutual sustainability – harmony rather than competition – and with the human participants benefiting more from the presence of wildlife than from its absence. I wouldn't know where to start addressing this.'

Four of the five contemporary migrations, including the Mara-Loita, are severely threatened and have virtually collapsed. The exception is the Mara-Serengeti, home to one of the last great land migrations of mammals on Earth. Since 1977, this has remained at around 1.3 million wildebeest, 200,000 zebra and 400,000 gazelle. However, the migration now spends less time in the Greater Mara Ecosystem each dry season; 2.5 months on average rather than four months (June through to the end of October).

Previously, up to a million animals moved into this area from the Serengeti during the dry season, but in September 2021, the number had dropped to just over 100,000. This is not a decline in the overall population, just that the migration must find food and water elsewhere.

Changes in land use and the climate, and the impact of large numbers of livestock grazing in and outside the reserve are the most obvious reasons for this downward trend. The future of the great migration is reason for real concern, given that the offtake for bush meat already accounts for up

Maasai Mara, Kenya. © Jonathan and Angela Scott

Maasai Mara, Kenya. © Jonathan and Angela Scott

to 140,000 wildebeest each year. In this regard, Niels Mogensen (Mara Predator Conservation Programme) noted: 'Lions rely on the wildebeest migration, and if it reduces in number or stops completely, lions are likely to suffer and reduce the number of resident prey substantially. We recently published a paper on the effects of variation in prey availability on prey switching in cheetah and lion.'[2]

The Mara River

Another major worry is the change in the health of the transboundary Mara River, which poses a major challenge for water resource managers and stakeholders. The river rises in the Mau Forest to the north, winding its way through the Maasai Mara and into the Serengeti in Tanzania before emptying into Lake Victoria. It's a vital source of water for domestic households, irrigation and livestock, and the lifeblood of the Mara-Serengeti ecosystem, essential to both resident and migratory wildlife.

In recent years, the river dried up in the driest times of the year from a combination of factors: deforestation and settlement in the Mau and the off-take of water by agriculturalists, as well as to generate hydroelectric power (the High Falls hydro-generating station project).

The river is increasingly being polluted by toxic chemicals from industry and raw sewage from the burgeoning human population. A recent die-off of fish in the Tanzanian section of the river with high levels of pollutants led a government official to ban the drinking of water from the Mara. If the river were to dry up, it could trigger an ecological disaster with grave consequences for wildlife, people and the economies of both Kenya and Tanzania.

In 2014, Tanzania gave notice of its intentions to implement a controversial plan to use Lake Victoria as an alternative water source for animals in the Serengeti National Park. This comes after decades of government efforts to help wildlife cope with the country's increasingly intense droughts. It involves reviving a 36km² wildlife corridor by extending the border of the park to Lake Victoria's Speke Gulf, to ensure the survival of millions of animals, including the great migration of wildebeest and zebra. But guaranteeing animals safe passage to the second-largest freshwater lake in the world will mean evicting hundreds of families living on the land. To date, this has not happened but has received all the necessary state approvals and is at an advanced stage.

Marsh Pride territory today

In the past 45 years, the Marsh Pride's territory has become much more open, with less bush and trees owing to drier times precipitated by the global climate crisis and some 2,500 elephants roaming the Greater Mara ecosystem. This has caused a decline in the riverine forest on the western boundary of the pride's territory. Elephants scouring the plains for seedlings and ripping them out to eat has prevented the regrowth of acacia bushes. The loss of cover along the intermittent watercourses – in particular, the Bila Shaka lugga – has been compounded by safari vehicles opening the croton thickets and thorn bush by forcing a way through them in their search for lions and leopards. Pastoralists have also had an impact by cutting down the bushes to help fence off their livestock enclosures. Opening the thickets allows seasonal fires to more easily burn into the heart of the wooded areas and further denude them. The loss of suitable cover has seriously compromised the ability of lionesses to find secure den sites for hiding newborn cubs to protect them from predators (and herds of buffalo) and isolate them from pride members until they have imprinted on their mother's scent and voice.

The spotted hyena population

Competition with the burgeoning spotted hyena population has further complicated life for the lions, with increased mortality of cubs. Lionesses start bringing their young to kills from the age of eight weeks. Cubs are still incredibly vulnerable at this age and smaller groups of lionesses struggle to keep them safe in these situations.

However, hyenas are afraid of the big male lions, another reason why it is so vital to have pride males close at hand when cubs are small. Hyenas, by comparison, suckle their cubs for more than a year, leaving them at secure communal den sites when they go out to hunt, avoiding potential conflict with other predators at kills.

Maasai Mara, Kenya. © Jonathan and Angela Scott

Eight years after the poisoning, it is time to act. What a miracle it would be if the demise of the Marsh Pride became the catalyst for serious dialogue and change, leading to full implementation of the recently ratified Management Plan for the whole reserve, embracing a strictly enforced embargo on grazing of livestock within the reserve and a moratorium on any further tourism development.

KENYA

13

Lions return to Amboseli

Limiting the reasons for lion hunting to either revenge for predation or as a quest for glory is to oversimplify it. Lions have also been killed as political statements against conservation policies.

Jeremy Goss

The small group of Maasai warriors have gone in search of a cow. She lagged on the walk home from a day out grazing and disappeared into the wilderness of the Amboseli Ecosystem. Eventually they find her but they are too late. The lions that killed her have left almost nothing behind. One of the warriors starts to shake. This was one of only three cows owned by his family. Another is her newborn calf, which will almost certainly die without its mother. The young man falls to the ground, the emotion pulsing through his body in convulsions. His companions rush to grab his spear and hold him still until the state passes.

This is it: decision time. The group is angry beyond words. They keep their spears sharp for exactly this situation. As warriors, their job is to protect their community's livestock, and a dead lion cannot kill again.

Twenty years ago, the lions would have been hunted down and speared to death. But instead, the emotional group turns for home to break the news to an equally emotional family. The question is why, and the answer holds lessons for the conservation of lions across Africa.

The Maasai relationship with lions stretches back as far as anyone can reasonably tell. As a group of people, they have maintained an intimacy with the species that has been lost across most of Africa, where the lion has long since been eradicated.

Unlike most human cultures, there was space for the lion in the traditional Maasai way of life as historically semi-nomadic pastoralists. The Maasai inhabit large areas of Kenya and Tanzania, much of which is wild, arid and not suitable for farming.

Livestock is central to their culture and livelihoods. The natural savanna habitats of 'Maasailand' provide grazing for livestock and a home for some of Africa's richest wildlife populations. In the past, there were no fences, few boundaries and little distinction between the human world and the natural one. Where humans walked, lions walked too.

Like so many relationships between man and animal, the connection between Maasai and lion is complicated. The lion is simultaneously revered and resented. Admired as a symbol of strength but loathed as a killer of cattle. What better adversary, then, against which to test oneself.

> The turnaround in the lion population is all the more remarkable in that it has been achieved on community-owned land outside of a national park or reserve.

Mount Kilimanjaro viewed from Amboseli National Park, Kenya.

Community members gather at the scene where one of three cows was killed by lions in one evening. The group is angry, but there is no retaliatory hunt despite the large financial loss.

For the Maasai, lion hunting has traditionally been elevated above the practical need to protect livestock. The killing of a lion was the pinnacle of achievement for a young man going through the 'warriorhood' period of his life. During this time, from roughly mid-teens to mid-twenties, the warriors or 'morans' were the fighting force of the tribe, responsible for protecting their communities and all-important livestock from other tribes and predators alike.

To hunt a lion was (and still is, in many places) a demonstration of the bravery and hunting skill not only of the individual, but of the warrior generation of which he was part. The hunts were planned in secret and spears were the only weapons allowed. After gathering at a predetermined location, the warriors tracked the lion/s and hunted as a group, but the first to draw blood in a successful hunt was recognised as the 'owner' of the lion.

The paws, tail and mane were carried back to the village as proof of the feat, putting the entire hunting party at the centre of local attention. The owner of the lion was celebrated with a lion name, and new songs and chants were created to memorialise the achievement. As each warrior age group reached the end of their tenure and transitioned to the next stage of their lives, their total number of lion kills was tallied for comparison with the generations gone before.

In contrast to this culturally driven lion hunting, other hunts were more spontaneous, usually a response to livestock depredation. The bond between Maasai and cow is certainly emotional – for many it touches on the spiritual – and the killing of a cow by a lion is a grievous affront that evokes intense anger in both the owner and community.

Further, as Maasai communities have integrated into cash-based economies, the monetary value of livestock losses to wild predators has become increasingly important. This value might range from $200 to $2,000, any of which is material, particularly for families who may have few other assets. Simply, lions cost people money they can ill afford to lose.

This lioness was killed by a Maasai hunting party after breaking into a boma overnight and killing several livestock. This retaliatory killing took place in an area where there is no Predator Compensation Fund to reimburse villagers for their livestock losses.

Participation in retaliatory hunts was not restricted to warriors but the method (spears), goal (dead lion) and community reaction to success (huge celebration) remained the same.

Limiting the reasons for lion hunting to either revenge for predation or as a quest for glory is to oversimplify it. A single hunt might be driven by multiple motivations and each participant may have a different reason for being there. More recently, lions have been killed to make political statements, often against conservation policies viewed as unfair or exclusionary to local communities.

Regardless of the motivation, the outcome of a lion hunt was always unpredictable and there were deaths and injuries on both sides. In a time of fewer humans and larger wild spaces, lion reproductive rates held up against the persecution by spears in Maasailand. Lions persisted.

A terrible discovery in the 1990s was set to tip the scales. Someone worked out that a weapon of mass destruction, originally designed to kill the smallest of animals, was equally effective at killing one of the biggest.

Carbofuran is a neurotoxic chemical, deadly to a wide range of insects that suffer the misfortune of eating the same plants that humans do, thereby earning the label of 'pests'. It turns out to be equally deadly to most animals, in scarily small quantities. Just one-quarter of a teaspoon is supposedly enough to kill a human.

A weapon as powerful as this was never going to be used strictly according to the instruction manual, nor solely on its intended target. It was cheap and available over the counter across most of rural Kenya and it quickly became the go-to for killing lions.

Gone was the need for tracking acumen, for bravery and skill with a spear. Now anyone could be a lion killer. In the moment of anger following the killing of one of your livestock, all it took was a sprinkling of carbofuran on the carcass.

With that small effort, you could kill not one lion but an entire pride. Better still, you would likely also kill some jackals and maybe a hyena or two, all of which are equally despised as livestock killers. Vultures would die, but that wasn't something to be worried about. There was no glory in the act and no celebrations or singing that followed. The killer would act in secret and usually alone, not doing it for fame but to eradicate an animal considered a dangerous and costly pest. The Maasai, as a community, view the use of poison as a cowardly act and it is frowned upon by the majority. Yet a handful of cowards can do a huge amount of damage.

In the Amboseli Ecosystem, the combined impact of spears and poison was devastating. Dead lions were stacking up much quicker than live ones were being born: 108 lion killings were confirmed between January 2001 and April 2006, but this figure may have been as high as 140.[1]

The lion population began to crash. No-one knows how many were left at the lowest ebb but as researcher Laurence Frank wrote at the time, lions were being killed 'at a rate which will ensure local extinction within a very few years'.

Prides had been decimated, the survivors lived in the shadows, solo or in small groups. Nothing less than a large-scale change in human behaviour was going to save the Amboseli lion population.

Richard Bonham and Tom Hill, of what was then the Maasailand Preservation Trust (now Big Life Foundation), knew this and their approach was simple: to ask the people killing lions what it would take for them to stop.

And so started a long series of discussions under trees. When discussing a topic as important as lion killing, everyone had to have their say, and everyone wanted to be heard. One thing kept coming up. If there was a way to offset the economic losses of livestock predation, there was a discussion to be had.

What transpired was the Predator Compensation Fund (PCF), an idea that was seeded by the affected Maasai community and co-created with Big Life Foundation (also known as just Big Life).

The idea of replacing the monetary value of livestock killed by predators was not new and is broadly known as 'compensation'. Yet, this is where the similarities between real-life intentions and the resultant programmes tend to differ so fundamentally in their design that they should be classed as entirely different concepts.

In this case, the foundation of the programme is a contract, signed by Big Life and the community leadership. That agreement has evolved over 20 years and now contains 27 clauses that stipulate everything from how compensation values are determined, to penalties for poor livestock husbandry, how disputes are resolved, and what fines and consequences are in place should the agreement be violated.

The mechanics are sophisticated. When a livestock animal is killed by a wild predator, it is the responsibility of the owner to secure the scene and call Big Life. Big Life dispatches a team of two verification officers (VO) on a motorbike: one is from the community and the other is an outsider, to reduce the opportunity for collusion and fraud.

The VOs examine the site and look at evidence such as tracks, how the livestock animal was killed and how much has been consumed, and piece together what happened as best they can. The compensation value varies based on the predator responsible, the type of livestock animal, and the circumstances of the kill.

So as not to incentivise poor livestock husbandry, the compensation value is reduced through penalties if there was negligence on behalf of the owner or herder, either through poorly constructed night corral fences, or livestock killed while lost and unattended. The no-penalty compensation value is set to be reasonable and is adjusted periodically, but on average it lags market price to reduce the incentive for people to allow their animals to be killed intentionally.

Once the circumstances and compensation value are determined, the livestock owner is given a credit note – a piece of paper recording these details – and this note can be redeemed for cash at the next PCF payday.

This is a mutual agreement dependent upon both parties having made commitments. Big Life funds the costs of running the programme and most of the compensation amount. The community contributes 30% of the cost of the compensation claims and has agreed not to kill predators in retaliation for livestock depredations. A violation of this agreement leads to significant penalties, both for the community and the individual or individuals responsible.

The agreement is as follows. Each ranch (generally covering hundreds of thousands of hectares) is broken into smaller zones, within which everyone tends to know one another. PCF paydays are held every two months of the year and the number of credit notes accumulates during each two-month cycle.

Should someone kill a predator during any two-month window, all unpaid credit notes for that zone are invalidated. The whole zone loses because of the actions of possibly only one person and this creates collective community pressure against retaliatory predator killing. In order to restart payments, those responsible must be identified and they are liable for a fine of seven cows per lion killed, as well as any government-enforced legal consequences.

The concept, though designed with community leaders, was based on unaccustomed ideas and, initially, community outreach was important in overcoming this: a short play and film were developed and teamed with repeated meetings and engagement with the community. Regardless, the early years of implementation were sometimes turbulent as the agreement was tested.

In the original iteration – in accordance with the focus on trying to reduce lion killing – compensation was paid only for cows killed by lions. But it turned out that cows killed by lions accounted for just a small proportion of overall livestock losses, between 10% and 15%. As later became clear, most predation cases involve hyenas killing sheep and goats.

While a spear targets a particular victim, poison does not. To fully address retaliatory poisoning against species other than lions, which often resulted in the deaths of multiple species (lions included), the agreement was expanded to cover all livestock killed by all predators.

The impact on lion killing was immediate. The programme was initiated on 1 April 2003 on the approximately 134,000-hectare Mbirikani Group Ranch, where at least 31 lions had been killed in the previous one and a half years. Since then, in the 20 years until March 2023, just 13 lions were killed on Mbirikani. This is a 97% reduction in the average number of lions killed on a monthly basis.

In 2008, the programme expanded to the 160,000-hectare Olgulului Group Ranch, where the impact was similar. At least 69 lions had been killed between January 2002 and the start of PCF in August 2008, at an average rate of more than 10 deaths per year. In the following 10 years, only 12 were killed, at a rate of 1.2 lion deaths per year.

One reasonable argument for the reduction is that there were no lions left to kill, and this was certainly a factor in the early years. However, monitoring by the science-based non-profit Lion Guardians has recorded a dramatic increase in the ecosystem lion population in the last two decades. By 2019, Lion Guardians were monitoring 252 lions (of all ages) across the ecosystem, equivalent to a seven-fold increase since 2004.

Human behaviour is extremely complex, and it is challenging to provide definite explanations for these changes. It would not be possible nor right to attribute all of the lion conservation success in the ecosystem to the existence of PCF. Complementary predator conservation interventions by Lion Guardians and Born Free Foundation have certainly contributed, as have general conservation programmes implemented by numerous entities.

In addition, PCF is effective only if predator killing can be detected, and the rules enforced as a result. In Amboseli, Big Life is able to do this through an extensive community informant network and large-scale geographic coverage of patrols by community rangers.

When PCF began, lion-killing behaviour changed, that much is clear. The following principles are likely to have been key to PCF's role as a core driver of this widespread impact:

1. Strong community leaders played a central role from the outset: not everyone wanted PCF, and many would rather have killed all lions to prevent further losses. Yet, after community leadership structures decided lion killing should stop, something had to be offered in exchange. PCF was this alternative, which was ultimately acceptable to the majority and endorsed as an agreement that applied to all.
2. Compensation addresses economic self-interest: it is unreasonable to expect someone to suffer continued economic losses and not take action. Large predators have been exterminated by humans across most of the world because of

this. PCF is a way of consoling an individual who has suffered a loss, at the time when they are most likely to retaliate in response.
3. It is available to all: much has been made of the need to raise the value of predators through economic incentives (e.g. performance payments or ecotourism revenue). These are important but often benefit only a subset of the community. Compensation is available to anyone who suffers a loss and is paid directly to the livestock owner; there are no middlemen.
4. Transparent and effective management: a transparent programme – fairly implemented and properly managed over an extended period – has built community trust. In the case of disputes, these are settled by a community committee.

The vast majority of livestock deaths are reported to receive compensation, and the resultant long-term dataset provides valuable insight into the nature of livestock predation in the Greater Amboseli Ecosystem and the severity of the cumulative impact on a community that shares space with predators.

In the almost 20 years between 1 April 2003 and 31 December 2022, Big Life compensated for 48,648 livestock killed by wild predators in 31,617 separate incidents, through payments totalling KSh160,784,555 (equivalent to at least US$1.6 million).

The area of compensation coverage varied over this period, starting at 134,000 hectares, before growing to 293,000 hectares, and reduced to some 220,000 hectares by the end of 2022.

Lions were responsible for just 9.6% of all livestock deaths, fewer than half of which involved cows. Hyenas were by far the biggest killer, responsible for 59.1% of all livestock predation: most were small stock (sheep and goats), while cheetahs were responsible for 15.5%, jackals for 13.1% and leopards for 2.1%.

As a telling indication of the most common reason for livestock losses – poor herding practices – the majority of incidents (60%) were judged to have occurred while livestock were lost or unattended.

There is no exit plan for PCF. It is an active solution to an ongoing problem and has to be maintained in perpetuity to have the desired impact. The running costs of the programme exceed the annual compensation amount, which is a necessary distortion given the importance of a well-managed programme. The three largest costs – which together account for more than 85% of the total – are compensation payments (~37%); salaries for 34 staff (~27%); and running costs for verification motorbikes (~22%).

In 2022, the total cost of the programme over an area of about 220,000 hectares was approximately $370,000, equivalent to just under $0.18 per hectare. Considering that this achieves blanket protection for all predator species, against all forms of killing, this is arguably a highly cost-efficient intervention.

Lion killing has not been eradicated, and violations still occur, the most noteworthy being the killing of six lions in May 2023 on Mbirikani. Incidents like this are definite setbacks but are frequently catalysts for much-needed dialogue, and the community-enforced penalties are important demonstrations of commitment to the agreement.

Given the strong recovery of the lion population in Amboseli, limits are now being tested as more lions spend more time in close proximity to humans. Additional work is required, particularly on improving herding methods and efficacy. Each avoided predation reduces the cost of compensation, and the number of livestock animals killed each year should ideally be reduced to an absolute minimum.

However, it is not possible to reduce predation to zero. If economic losses are not balanced at the household level, it will be extremely difficult to conserve lions in large parts of their current range, which are outside of formal government-protected areas and where lions come into contact with humans and their livestock.

In Amboseli, PCF has demonstrated that a community-initiated compensation programme, effectively and consistently implemented, can have extraordinary success in reducing lion killings. The turnaround in the lion population is all the more remarkable in that it has been achieved on community-owned land outside of a national park or reserve.

Local adaptations would be necessary to implement the PCF concept elsewhere but the approach, focusing on the eonomic requirements and realities of people whose actions are leading to wildlife population declines, is something that can and should be successful in any circumstances.

Lion in the Chyulu Hills, which form the natural corridor between Amboseli and Tsavo West national parks, Kenya.

KENYA

14

Lions on the doorstep

Nairobi National Park is right up against a densely populated city, a situation to which both lions and humans need to adapt.

Emma Childs

We have lived in different lion territories since 1997, all of which except one have bordered Maasai community land. The most extraordinary population of lions I have ever come across are those of Nairobi National Park. On one side, there is a densely populated metropolis with humanity right up against the borders, and on the other, an open area where for decades lions have moved freely through the southern conservation area: Nairobi National Park.

In each area we have lived, we've seen normal, standard lion behaviour. Of course, human–wildlife conflict has been rising and for the last few decades, lion populations have been on a downward trajectory. The Nairobi Park lions, however, are different in how they behave. Living alongside them on the border with the park over the past 11 years, I have been amazed at their level of adaptability.

Nairobi National Park was established in 1946 and is Kenya's oldest park. Today, it's fenced on three sides and the area where I live is an open boundary, the Mbagathi River being the line between community and wildlife.

When conservationist Mervyn Cowie, who was born in Nairobi, returned to Kenya in 1932 after a nine-year absence, he was alarmed to see the attrition of game animals on the Athi plains.[1] Expanding farms and livestock had taken the place of the game. It was a paradise, he later recalled, that was quickly disappearing.

The area – which would later become Nairobi National Park – was part of the Southern Game Reserve. Hunting was not permitted, but nearly every other activity, including cattle grazing, dumping and even bombing by the Royal Air Force was allowed. Cowie campaigned for the establishment of a national park system in Kenya and the government formed a committee to examine the matter.

Nairobi National Park was officially opened in 1946 – Kenya's first national park – and Maasai pastoralists were removed from the area.[2] Cowie was named as director and held this position until 1966. In 1989, Kenyan President Daniel arap Moi burnt 12 tonnes of ivory on a site within the park, hugely improving Kenya's conservation and wildlife protection image.[3]

When I was growing up in Nairobi, my father would take me into the park and on these occasions, we rarely saw any cats and never rhinos.

Nairobi National Park, Kenya. © Paras Chandaria

Nairobi National Park, Kenya. © Rihaz Sidi

In the 1980s, the park was part of a much greater ecosystem that stretched into Amboseli and Tsavo, but, over time, this corridor has been closed as people have bought up land. The result is that we now have a pool of lions that have become 'islandised'. Now and again new blood comes in but, generally, this population's gene pool is shrinking.

A 2021 management plan for the park showed there were around 45 lions in an area of 114km^2, with the population continuing to rise. In the 11 years of living on the border at The Emakoko lodge, we have seen the rise and fall of 'great kings', the rise and rise of female-dominated groups and the extraordinary behaviour of youngsters. There does not seem to be the stand-alone pride of more typical ecosystems, where one male dominates a female-heavy group. Here, the population consists of female groups with males simply moving through.

My husband Anthony decided a lodge on the border of the national park would be a good idea and we hunted down a likely area for such a venture. Anthony camped for several nights along the border in the Maasai community zones until he found a spot that would tick all the boxes, including being off the flight path but with easy access to the main airport. In those days of camping, he spent many nights speaking to the community about conflict with wildlife. However, in the areas where he camped there appeared to be little, if any, wildlife. In most places, the nights were silent, then everything seemed to kick off between midnight and 04:00. Lion roars would come down the valley, no doubt accentuated by the valley acoustics. Occasionally, a stray lion would wander through and cause havoc with livestock, but this was relatively rare.

When we were putting the final touches to the lodge about a week before opening, we had our first encounter with lions. A building crew were cladding the lodge fireplaces and working day and night to finish on time. A couple of middle-aged lionesses decided to investigate, crossed the river and walked right into the lodge. Two builders came face to face with them in what is now the lounge area and hastily fled to join the rest of the crew, who were in other rooms. We all went back to investigate as it seemed unbelievable that two lions would just walk into a building full of people. Yet, there they sat, completely unperturbed. We decided to leave the evening to them and packed up. The Maasai night guard said they spent a good hour or two in the main areas the following morning before crossing back into the park.

I had never before seen lions behave like that, but it appeared to be the norm in the area. The lions were absent during the day, but at night it was a different story – on many occasions I have chased off lions as I walk back to my house.

At one point, visits by lions became intense. We'd had a month of endless rain, and a lioness in the lodge grounds became marooned, with no way of getting back over the river into the park. We always slept with our doors open, our dogs serving as an early warning and deterrent. Late one night, the stranded lioness came right up to our house, probably starving and attracted by our baby who had been crying. Our ferocious Jack Russell, which slept at the end of my bed, was not going to take the intrusion lying down. She leapt out of the door into the jaws of the lioness, who was only about 30cm away. Sadly, our dog did not survive. We assumed the intruder was a leopard until we found the tracks the next morning. Over the next week, several dogs from the area were taken. Anthony tracked the lioness to where she had been hanging out just below the staff camp. Once the river receded, she made her way back into the park, but not before taking another two of our dogs. It was a hard lesson about the inappropriateness of keeping domestic animals in wildlife territory.

Perhaps having been squeezed out of a female pride, the lioness didn't go far, making her home in the park just across the river, perhaps having acquired a taste for domestic animals. She later had cubs, which acquired the same taste, venturing onto community land and taking more than just the dogs. The lioness had to be captured and relocated to Tsavo East National Park.

Generally, though, human–wildlife conflict between lions and the community has been relatively limited. The rains have not been great, which means our big cats have had an easy life taking down drought-weakened prey. Conflict tends to happen when the community bring cattle into the park. Even bigger problems occur when the rains are good and the game moves out into the community land, followed by lions. Considering this, the community has been extremely tolerant, perhaps understanding that they and the lions share the park.

More recently, however, lions have been escaping the park in the other direction, perhaps because of internal conflict or maybe just curiosity. A large male lion was seen running through traffic along the Mombasa Road during rush hour, much to the horror of pedestrians. In the quiet, leafy suburb of Mukoma, a lioness ventured into gardens on more than one occasion until she, too, was darted and relocated to Meru National Park.

Over the years, our guides and guests have witnessed the rise and fall of dynasties. In 2021, we had our greatest loss when the magnificent Sirikoi had to be destroyed after shattering his jaw trying to get out of the cage while being taken to have an injury fixed. He was the son of the great Mohawk. They were two of the 'great' lions of this park that seemed to have ruled the area since the early 2000s.

As of 2023, we have six large male lions. They usually traverse the park using the roads and do not move aside for anyone. Two dominant males – Mpakasi who is just over 10 and Kitili who is about nine – have formed a coalition. Another coalition on the northern side of the park is between younger brothers Leshan and Quintai, and will most probably be the next force to be reckoned with when the Mpakasi kingdom falls. There's a new and very young coalition on the southern side, and also two sons of Kitili, Elengat and Selengai, almost four years of age.

For the moment, though, the park belongs to Mpakasi and Kitili. They have ruled – in relative peace – since the death of Sirikoi. But trouble could loom up ahead from Elengat and Selengai as they seem to have closed off their area to the kings. In a fierce showdown, Mpakasi and Kitili were booted out of the southern area and chased back into the middle of the park. Will they continue to rule or will they fall? Despite Leshan and Quintai being next in line, they appear to have no interest in challenging anyone. As long as there is food and water, they have no loyalties to anyone and do not care who passes through their turf.

The largest pride is female dominated and lives around the lodge. There are two dominant females: Solo, who at 14 is one of the oldest lionesses in the park, and her niece Nala. They are a highly successful group, which includes six cubs. This pride is well known to the community around The Emakoko as they venture into the community land. However, they are well behaved and have no taste for livestock. From time to time we bump into these ladies in the lodge car park. Another nearby group consists of two females with three cubs, all of which seem shy and keep out of the way of Solo and out of the community.

The most commonly seen group of lions in the park are the Mpakasi and Kitili prides, consisting of four large females and 12 cubs of different age sets, ranging from three to eight months. They pose no threat to the community and tend to hang out in the 'central district' where the two kings frequently prowl. A few years ago, there was a bitter spat in the group when one of the youngsters decided to chat up the ladies. It quickly all went wrong. He was pounced on ferociously and had to scuttle off, tail between legs, gashed and bruised.

Towards the end of 2022, the drought hit the area very hard and rains that should have arrived in October finally fell at the start of December. The community had lost most of their cattle and whatever was left was moved out by truck to areas where there was grazing. The park was all that was left in the area with grass and, although dry, it was able to keep our migrant herbivores alive. By the end of the year, the buffalo herds had become large, some of them over 100 strong. Eland, impala, zebra and wildebeest grouped in large numbers in areas that contained the last vestiges of grazing. This abundant wildlife offered easy pickings for our feline friends and there has been little to no cattle–wildlife conflict in the area.

There is now talk of fencing the park but the community is fighting it as, in dry seasons, the park allows them to graze their cattle there, even if it's illegal. However, wildlife services must provide compensation to the community for loss of livestock and it sees the cost of fencing the park to be the cheapest solution. A fence would also prevent wildlife from moving in and out of the park.

For more than 10 years we have been lucky enough to observe these incredible creatures from our refuge here at The Emakoko. Humanity has pressed in on every inch of the park, but these lions have adapted to the changing situation. Each generation that comes through seems to have updated their relationship. These big cats will continue to become more relaxed around people as long as they are not persecuted.

'What the Nairobi National Park lion gene pool and behaviour will look like in 10 years is anyone's guess. Time will tell. One thing is for sure, as lion populations go, the resident prides are unique and need our protection.'

Emma Childs

Nairobi National Park, Kenya. © Paras Chandaria

KENYA

15

Lions, livestock and extinction

Lions eking out an existence between parks are in deep trouble.

Dr Laurence Frank

When did you last see a grizzly bear in California, a wolf in New England, or a tiger in Türkiye? All were present just over a century ago, and all have been shot, trapped and poisoned to local extinction. Lions in Africa now face the same fate.

Ten thousand years ago, lions had the widest geographic range of any mammal, and were found across Europe, Asia, Africa, North America and the northern half of South America. Today they occur only in drastically reduced range and numbers in Africa, with a few hundred in Gujarat, India.

Large carnivores were the first animals to go extinct when modern firearms and poisons arrived in undeveloped regions; and they are the most difficult animals to conserve in the modern world, in part because they are a threat to human life but primarily because of their ecology – preying on ungulates, large grazing animals. Humans are extravagantly fond of some ungulates: the cattle, camels, sheep and goats we domesticated perhaps 12,000 years ago. Compared to alert and wary wild grazers, domestic livestock is abundant and easily killed, particularly where wild prey has been depleted through habitat loss, poaching and overgrazing by cattle, sheep and goats, as is the case in much of today's Africa.

As a young man in 1970s Kenya, I was accustomed to seeing lion tracks and hearing their roars at night in much of the country, sometimes far from national parks. Since then, the human population has exploded five-fold and wild lands have been ploughed under for farming or overgrazed by livestock. Wildlife has disappeared and the nights have turned silent – the wild grazers snared as bushmeat, and the predators speared or poisoned for killing cattle.

Vast areas where once there were lions now support no wild mammals larger than hares. The decline was obvious to local people, but the world at large remained unaware because, in national parks, lions seem to lounge under every other bush, surrounded by minivans full of excited tourists. Only when Philippe Chardonnet, Sarel van der Merwe and Hans Bauer published papers in 2002 and 2004 documenting the dramatic Africa-wide decline did the world become aware that lions had largely disappeared outside of protected areas.

In the past, large carnivores posed a threat to human life; the last thing seen by a great many of our distant hominid ancestors was a charging lion or lungeing hyena. However, with the exception

A traditional Maasai manyatta near Chyulu Hills National Park serves as a home for a Maasai family and their livestock. Given the presence of free-ranging lions in the area, these manyattas are built with a strong focus on security, incorporating thorny fences around each homestead to safeguard against wild animals.

of a few small areas – southern Tanzania and the Sundarbans swamp of Bangladesh – human fatalities are extremely rare today. Rather, it is the depredation of livestock by lions and other large predators that has caused their catastrophic worldwide decline.

Before the advent of modern technology, humans developed a variety of methods for protecting their animals from predators, but those involved large investments of time and energy. Traditional methods of livestock protection in Africa differ significantly from those of Europe and North America.

The primary economy of African pastoralists – formerly nomadic but now largely settled – is based on their livestock. By day, young men armed with spears, short swords and clubs accompany their grazing herds, leading them to fresh grass and water, and then herd them back to their settlement before dark. Livestock is held overnight in corrals (called bomas in East Africa, and kraals in the south) built of dense thornbush, with people sleeping in huts around them and dogs to warn of approaching danger – a practice requiring intensive and dedicated manpower.

This type of system must have been developed early in the domestication of grazers, a time when large carnivores were ubiquitous. The grand experiment in domestication would have failed immediately had humans not developed effective ways to protect their animals from predators – and from other people.

Once pastoralism became established, so did stealing livestock. Still today, cattle raiding is a way of life for young men among the herding tribes of northern Kenya and elsewhere in Africa where pastoralism still survives. Thus, guarding their fathers' cattle is the primary activity for warriors, when they are not off raiding cattle to build their own herds, without which they cannot afford the bride price for getting married. AK-47s make raiding today a far more lethal sport than it was in the days of spears and clubs.

In western Europe, wolves and brown bears had been largely eliminated well before the 20th century and humans no doubt played a decisive role in the much earlier Pleistocene extinction of lions, spotted hyenas, cave bears and perhaps even sabretooths in Eurasia and North America.

When Europeans arrived in North America, they gradually exterminated wolves and pumas as they moved westwards and, once they crossed the Mississippi, modern firearms, traps and poisons massively reduced wolves and grizzlies within a few decades. White settlers first arrived in California in large numbers only in 1849, during the great Gold Rush, encountering the highest density of grizzly bears known anywhere, and shot the last one just 73 years later.

Without great predators (or human rustlers), livestock can be simply turned loose to graze unprotected, freeing up their owners for other pursuits. In the absence of predators, the modern western beef industry requires relatively few people. Although wolves and grizzlies are absent except in the northern Rockies, many pickup trucks in ranching country still sport a rifle rack in case of an opportunity to shoot a coyote or bobcat. Aside from state and private predator control, the US government kills in the order of 75,000 mammalian predators annually. In 2022, Idaho passed a law to allow the killing of 90% of its recently restored wolf population. In western Europe on the other hand, wolves and brown bears are increasing and expanding their range thanks to public support for wildlife conservation.

Today, we are witnessing a similar collapse of large carnivores in Africa. Although European settlers had decimated wildlife in farming regions, lions, hyenas and their wild prey persisted on much of the continent until the late 20th century. Lions remain abundant where they are protected but have largely disappeared from the vast unprotected rangelands that separate the few well-protected parks, most too small to ensure long-term viability of their lion populations.

One of the greatest problems facing wildlife today is that human development disrupts ecosystems by dividing them into fragments isolated from each other by agriculture, urban areas and roads. This fragmentation reduces the connectivity between protected areas and their wildlife populations that would allow animals, and their genes, to move between them.

Any number of circumstances can decimate an isolated population: disease, inbreeding owing to small population size, an influx of humans and livestock in response to social unrest or simple failure to enforce boundaries. Corridors are stretches of safe habitat between populations with sufficient food, water and cover. They maintain connectivity by allowing animals to move between safe areas, effectively creating a single larger population by linking several smaller ones.

For lions, this means maintaining viable breeding populations in the livestock rangelands between parks, which was the case until the late 20th century. To address predator conservation outside of and between parks, where livestock grazing is the primary land use, I established the Living with Lions organisation.

For a variety of socio-economic reasons, pastoralists in many areas had lost whatever tolerance of lions they had previously accorded them. In Kenya's Maasailand, the rate of spearing by young warriors increased dramatically around the year 2000 and in northern Kenya, relentless overgrazing and abundant cheap automatic weapons have taken a severe toll on all wildlife.

The universal availability of cheap and highly toxic agricultural pesticides made widespread indiscriminate predator poisoning a simple way to permanently eliminate livestock losses to lions and spotted hyenas. European farmers had earlier introduced the use of older poisons such as strychnine and acaricides (formerly used to control ticks on cattle, now replaced with less toxic compounds) to eradicate predators in farming regions and even in some wilderness areas.

In his career as a game warden, the famous conservationist George Adamson never travelled his northern Kenya beat without a can of strychnine in his Land Rover, with which he poisoned hyenas at every opportunity. Postcolonial government agencies continued this legacy, and at least one was still poisoning hyenas nearly into the current century.

Because they return to a large kill the following night, poisoning lions could hardly be simpler. The aggrieved cattle owner need only sprinkle the carcass with a few cents worth of pesticide to kill the culprits, as well as every scavenger that visits it. It is not uncommon to find several dead lions, along with hyenas, jackals and dozens of vultures next to a partially eaten carcass. Until very recently, six species of vultures and many different eagles – often Eurasian birds wintering in Africa – were abundant; a glance at the sky would always reveal large birds soaring high above, and after the sun rose, any fresh carcass was covered by squabbling vultures and made to disappear within a few hours.

Thanks to current widespread predator poisoning, all African vultures are rapidly declining towards extinction, and dead animals just dry out in the sun, untouched by scavengers.

It was the introduction of carbofuran, a crop pesticide originally made by an American company and sold under the brand name Furadan, that ravaged many lion populations, sometimes in remote areas. Carbofuran's toxicity to birds and mammals led to its being banned in Europe and North America, but in East Africa, every small shop catering to farmers stocked Furadan, and news of its efficacy in killing predators spread quickly among pastoralists.

Immediately following an exposé of its impact on lions and other wildlife by the American news programme *60 Minutes* in 2009, the manufacturer withdrew Furadan from the East African market and poisoning seemed to decrease, but pastoralists switched to other pesticides, including imported generic carbofuran, and in some areas, poisoning continues to be a major cause of lion mortality.

Ironically, when applied diligently, the ancient livestock management practices are still remarkably effective at protecting domestic animals from predators. Yet, herding livestock from dawn to dusk is a dull way for a modern young man to spend his days and, because thornbush decays, bomas require constant maintenance, usually the responsibility of women. Further, in many areas, the great increase in human and livestock numbers has drastically reduced the availability of trees and stout bushes for boma construction. Nomadism has given way to permanent settlements, where bush is used for firewood and commercial charcoal production, as well as for construction.

If one can eliminate lions and hyenas with spears, guns or poison (and if raiding has been suppressed, as is the case in Kenya's Maasailand) young men can spend their time in other pursuits. Children, unable to repel lions, are often assigned to herding. African women have enough hard labour without also having to get involved in boma building and maintenance – chores happily forgotten. In Botswana, herding has been largely abandoned and the cattle are left to graze unaccompanied, finding their own way back to their kraal or spending the night in the bush. Needless to say, unattended cattle are easy prey for any lions and hyenas that persist in these areas.

Lions typically take cattle at night by stealthily approaching a boma and causing the cattle to panic and stampede, bursting through the weakest point – likely a bush that serves as a gate when dragged across an opening in the boma wall. In recent years, the deficiencies of thornbush bomas have been addressed through the use of chainlink fencing in different configurations. Laly Lichtenfeld and her colleagues in northern Tanzania's Maasailand have developed a highly effective low-maintenance 'lion-proof' permanent boma constructed of living walls – comprising fast-growing Commiphora trees – supporting a chainlink fence, at a cost of roughly US$500 per boma.

In Laikipia County, Kenya, many commercial beef ranches have small tourism operations supporting robust lion populations of about six adults and subadults per 100km^2. Nearly all have adopted lion-proof mobile bomas developed by local rancher Giles Prettejohn, who modified and scaled up an earlier design for sheep developed by John Harris. These consist of interlocking panels of welded water pipe supporting chainlink. At about two metres in height, these, too, are nearly 100% effective in containing panicked cattle and preventing lions from jumping in. They are readily dismantled and moved across the landscape every few months to follow grass and water availability, removing the need for permanent bush bomas, the construction of which requires cutting of trees and bush.

They have the additional benefit of regenerating soils damaged by overgrazing in the last century, as once the boma is moved and rain falls, the accumulation of dung left behind nourishes lush grass and the gradual development of new soil. However, they cost US$2,000–US$3,000 each, and require a tractor or truck to move them, putting them beyond the reach of most pastoralists. To make chainlink construction feasible for pastoralists, non-governmental organisations or government subsidiaries are needed; and, for mobile bomas, maintenance and repositioning would require a technician and a small truck.

While both ancient and modern boma designs effectively protect livestock at night, neither are of any use if people are not motivated to construct and maintain them. It will always be far simpler, cheaper and longer-lasting to just eliminate predators and turn livestock out to graze 24 hours per day, 52 weeks per year.

Lions will not survive outside of protected areas unless the people on those lands make the necessary efforts to protect their livestock, rather than simply eradicating large carnivores as has happened in other parts of the world. In most cases, that motivation must be financial; proceeds from tourism or other wildlife-based activity will provide enough income to local livestock owners to more than offset the cost and effort of herding and boma construction.

Even tourism is not a universal solution. It is concentrated in scenic areas with infrastructure to provide comfort and safety for international visitors, while the most remaining lion range is often of low tourism potential: hot, remote and covered in monotonous bush, and wildlife that is hard to see. In areas with substantial tourism, little of the income may reach the pastoralists who bear the costs of predators. Most of the money stays with tourism operators and governments, and although some filters down to the local community level, it frequently enriches only a few influential individuals rather than the little guy for whom the loss of a single cow is a major financial and emotional hardship.

One excellent counterexample, however, is the conservancies north of Kenya's Maasai Mara Reserve, where tourism operators have helped Maasai landowners join their separate holdings to form protected areas where grazing is controlled and wildlife is protected. In return, the landowners earn an excellent income from upscale lodges in their conservancy. Not many years ago, lions were heavily persecuted in these lands and few survived. Since the establishment of the conservancies, they are now abundant.

Another novel approach to lion conservation in pastoral landscapes is the Lion Guardians programme, developed by my colleagues Leela Hazzah, Stephanie Dolrenry and Seamus Maclennan as part of our Living with Lions project, which addresses lion conservation outside of protected areas in Kenya. The Guardians employ young warriors, most of them uneducated, illiterate and unemployable in the modern world, to assist their communities in avoiding cattle losses by improving bomas, helping with herding, and finding livestock lost in the bush. Most are former lion killers and know their lions, their communities and their region intimately.

Among the Maasai, killing a lion confers great prestige on a young man, and the Guardian programme maintains that cultural connection between lions and warriors, but through non-lethal means of protecting their community from depredation. They have also been central to research, able to assist biologists in gathering lion data on a regional scale, something that is otherwise difficult using standard wildlife research techniques. Given their status in their community, they are able to pacify other warriors angered by a lion attack on their cattle and intent on spearing the culprits. The Guardians have turned around a lion population, which was being speared and poisoned to extinction, and that success has led to the programme being replicated elsewhere in pastoral landscapes.

Of course, maintaining healthy numbers of wild prey is critical, and this requires control of bushmeat poaching and overgrazing. In many areas, bushmeat extraction has reduced wildlife to levels that can no longer support predators. Snares not only decimate wild grazers, but directly kill countless predators inadvertently caught in them. Overgrazing the land is another factor, generally reducing its capacity to support life. Neither of these problems have straightforward solutions. Antipoaching programmes require considerable labour and funding, as well as alternative sources of protein for burgeoning human populations. These, and the political will of governments, are all usually in short supply.

Overgrazing often stems from deeply rooted cultural values of pastoral people, for whom livestock represents wealth and prestige as well as sustenance. As a result, elders accumulate large herds of animals without participating in beef markets, sometimes keeping old animals rather than selling them once they have reached adult size.

Thus, the land must support many more livestock than in commercial systems where most animals are sold off upon reaching marketable age.

Constant heavy grazing was much less common when pastoralists and their herds were nomadic, but permanent settlement causes severe degradation of grasslands, and the soil washes away when it rains. Semi-arid northern Kenya and the Horn of Africa were once productive rangeland, supporting abundant wildlife as well as sustainable numbers of livestock. In the last half-century, however, severe overgrazing has reduced the land to rocky desert, the remaining grass cannot support the large numbers of livestock, and little wildlife remains.

If livestock were removed today, it would take thousands of years for the soil and grassland to regenerate. Maasailand in southern Kenya gets much more rain, but even there, unsupportable numbers of cattle have seriously damaged vast areas in recent decades. And when the inevitable drought hits, livestock die, people's wealth turns into desiccated carcasses, and starvation stalks the land. In earlier times, there was usually enough grass to see animals through drought, but today, modest dry periods decimate both livestock and any remaining wild grazers. While climate change is frighteningly real, the devastating effects of periodic drought today are, in fact, largely the result of land degraded by long-term overgrazing.

As with so many conservation problems, both the cause and the solutions lie with the people who own and use the land. While westerners may be deeply concerned about today's worldwide extinction crisis and willing to spend money to reverse it, land and wildlife management are up to local people; even when governments enact legislation to better conserve and manage natural resources, they frequently lack the means of enforcing it. In Kenya, the two areas in which lions have recovered after earlier decimation, the commercial ranches of Laikipia and the conservancies north of the Maasai Mara National Reserve, are both under private control and the residents profit from wildlife tourism.

UGANDA

16

Murchison Falls National Park

This park has so much potential, yet it is being ripped apart by poaching and greed.

Dr Paul Funston

Uganda's largest national park, Murchison Falls, has the rather dubious distinction of being known as the 'snaring capital of the world'. Research indicates that even in open countryside where there's high-density game viewing, there are, on average, five snares per square kilometre. Rural communities living outside the park have taken the snaring a step too far, and refer to the park as a 'wild bank'. They are 'eating' it to death.

Anywhere in the park with moderate to dense woodland is essentially devoid of wildlife. Populations have been obliterated – snared, caught in gin traps or speared for food. And yet, in one key area, wildlife abounds in such breath-taking numbers that tourists visiting the game-drive circuits can only be astounded by the sheer volume of plains game.

Hundreds of thousands of Uganda kob stream across the hills and valleys on their daily foraging rounds, and there are good numbers of Jackson's hartebeest, waterbuck, buffalo and giraffe – more than enough to keep the relentless stream of budget-safari tourists satisfied. Smaller herbivores abound too, with many warthogs and oribi to keep the leopard population well fed. But, although the habitat looks perfect for them, there are no cheetahs or African wild dogs in Murchison. Buffalo, the lion food of choice, are steeply in decline, while kob and other common prey species – at least in one area of the park – are increasing. With so many contradictions, it's easy to be mistaken in assessing how the park is placed to conserve lions.

All of the activity occurs on the open grassy plains in the northwest section of the park, which is easily accessed by a main tar road that branches off onto many gravel tracks. In the past, Murchison, along with Uganda's Queen Elizabeth National Park, supported the highest mammal biomass densities of any protected areas in Africa. Along with that were very healthy lion populations. Hundreds of lions, at densities of over 20 per 100km^2, made Queen Elizabeth and Murchison unmatched in Africa for the ease with which lions could be found and watched, typically munching away on hapless buffalo. The prides were exceptionally large: 30 lions at a time were not uncommon.

Today there are still some large prides in the areas to which tourists flock every day: the Delta, Pakuba and Borassus prides. But these prides are

144 ~ THE LAST LIONS

now dwindling in numbers, and seem to have done so rather steeply since COVID brought a halt to tourism in the park. The biologists tasked with tracking lion numbers are reliant on a method involving estimates that does not hold up when one extrapolates the results over large areas with few to no lions. With these potentially inflated figures comes a false narrative regarding how well Murchison's lions are faring; and the result is that alarm bells are not being raised.

It's not hard to fathom why lion numbers are in decline. There are other interests at play in Murchison. The rich soils that support the vast herds of grazers also contain significant stocks of shallow oil. Not only have the Ugandan government and park services failed to curtail rampant poaching but, in a highly controversial and much publicised move, have signed over much of the park to the oil-mining company TotalEnergies. This company is ripping up the very habitat in which the last remaining lions in the park live, and pumping large quantities of crude oil through a pipeline to the coast in Tanzania.

Never, in all my many visits to national parks in Africa, have I seen such utter dereliction of duty: this once majestic park is now beset by greed and poaching. The hundred or so remaining lions and their boundless prey populations just don't deserve this.

In spite of the desecration, the park still offers visitors a lingering glimpse of the very best of East Africa wildlife. If Murchison could be fixed, if the areas of the park that now lie fallow and moribund could be rehabilitated and restocked with game, Murchison could easily become one of the best national parks in Africa again – one in which tourists could see wildlife in a landscape that is truly awe inspiring in terms of the volumes of wildlife, thanks to the immensely fertile soils. And such a turnaround could support the surrounding communities through related enterprise.

For now, we hold onto the hope that a major NGO might partner with the government to restore Murchison Falls to its former glory. But this seems a distant vision – the work that is being done is not enough; the posturing by and between NGOs is counterproductive, and both the wildlife and the livelihoods of humans are suffering. As you read this, imagine a lion or a kob struggling in a vicious snare.

In the face of ongoing decline, we can only hope that the four or so remaining prides can hang in there.

Lions on the plains in Murchison Falls National Park, Uganda. © Dr Paul Funston

Murchison Falls National Park, Uganda. © Dr Paul Funston

UGANDA

17

Tree-climbing lions of Ishasha

Lions are terrestrial creatures with their feet firmly on the ground – except around Lake Edward in Uganda where they hang out up among the leaves. But why?

Dr Tutilo Mudumba

In the beautiful valleys between the Virunga volcanoes and the Rwenzori highlands lies a rich savanna landscape adorned with majestic acacia trees and framed by serene Lake Edward and rugged escarpments. Amid this iconic scenery along the Ishasha River, something remarkable occurs: the presence of lions, renowned for their extraordinary tree-climbing abilities. Members of this unique group spend a significant portion of their time perched among the branches, defying conventional expectations of what are generally known to be terrestrial predators.

The act of climbing trees is not exclusive to lions; it's a natural skill possessed by all members of the felidae family, facilitated by their retractable claws, muscular limbs and agile bodies. However, what sets the Ishasha lions apart is the frequency and duration with which they engage in this behaviour. While lions in other regions occasionally climb trees, this is often limited to specific individuals. The subpopulation of Ishasha, however, can frequently be found perched on tree limbs for several hours at a time. This begs the question: what compels these predators to spend their days up in the branches?

Safety, comfort, or both?

One prevailing theory suggests lions climb trees to seek respite from the sweltering heat, benefiting from the cooler temperatures and refreshing breezes found higher off the ground. This proposition finds support in the fact that Ishasha's tree-climbing lions live at elevations below 1,000m above sea level, where daytime temperatures can be as high as 35 degrees Celsius. Lake Manyara National Park, situated at 960m above sea level, and Ishasha, at 990m, are in the lower regions of the western arm of the eastern Rift Valley. In such conditions, it is plausible that lions climb trees to find comfort and perhaps evade biting flies, which are prevalent during the rainy season.

Observations made in Ngorongoro in the early 1960s indicated that lions climbed trees more frequently during periods of high fly density. However, the biting-fly theory fails to fully explain why the phenomenon of tree climbing persists beyond seasons of high fly density, including during the dry season, an anomaly also noted in Manyara in the late 1960s.

Kyambura Gorge on the eastern edge of Queen Elizabeth National Park, Uganda. © Paul Goldring

Queen Elizabeth National Park, Uganda. © Wild Horizons

Queen Elizabeth National Park, Uganda. © Wild Horizons

Another possibility that drives lions to seek refuge in trees is the potential threat posed by elephants and buffalo, which will attack and kill lions if they fail to escape swiftly. I have made several field observations of lions either fleeing such an attack or climbing trees to avoid being trampled. Slower or younger lions that cannot escape in time have fallen victim to stampedes. Areas with high densities of buffalo and elephants often witness such incidents, compelling lions to climb trees as a matter of life and death.

Ishasha, like Manyara, shares similar geographical features, including a lake and impassable terrain comprising escarpments and deep ravines. These natural barriers form narrow passages that migrating buffalo and elephants traverse, particularly during the dry season. In such circumstances, tree climbing becomes an indispensable survival strategy. Over generations, this behaviour may result in the selection of tree-climbing lions, culminating in their prevalence observed in Ishasha.

It is plausible that both safety and comfort contribute to the population's propensity for tree climbing. For instance, lions that climb trees to escape death or injury may find the branches more comfortable owing to reduced exposure to biting flies and cooler temperatures. As a result, they may linger in the trees long after the immediate danger has passed, eventually exhibiting the behaviour more generally, even in the absence of threats.

Conversely, if seeking cooler temperatures was the primary driving force, lions may discover that remaining in the shaded areas below the trees reduces their need to run around or involuntarily climb trees to escape threats. Consequently, individuals that do not climb would be selected against, leading to a higher prevalence of climbers within the population.

To seek safety, comfort or both, the availability of trees with shortish trunks and large branches is crucial to support a tree-climbing habit. In Ishasha, the trees most climbed are the figs (*Ficus sycomorus*). From my preliminary study of the lions there, I found that out of the 178 mature fig trees, lions have been recorded in more than 50%, compared to less than 0.001% of any other tree species. This indicates a strong preference for fig trees, although lions also climb *Albizia coriaria*, various acacia species and *Euphorbia candelabrum*, among others. However, at Lake Manyara, vegetation and tree climbing are not correlated, so tree structure was not a key factor or predictor of tree climbing in lions there.

In Ishasha, where lions can be observed grooming, urinating, sleeping, stalking, roaring and engaging in territorial disputes, it's evident that the habit of tree climbing does not impede their expected social and biological behaviours.

Despite the captivating nature of tree-climbing lions, the population within the Ishasha sector has never exceeded 40 individuals since studies began in the early 1980s. The most recent lion population estimate places their numbers at fewer than three individuals per 100km^2, which is a worryingly low figure for an iconic species that ranks second only to mountain gorillas as a reason for tourists to visit Uganda.

With ecotourism playing a vital role in Uganda's economy, it becomes crucial to preserve lion populations for the long-term sustainability of the industry. The good news is that spotting these tree-climbing lions in Ishasha is relatively easy compared to those hidden in bushes. The presence of popular fig trees, conveniently accessible along safari tracks, serves as helpful markers for both visitors and researchers to observe these majestic creatures with ease.

My research aims to deepen our understanding of tree-climbing behaviour among lions and explore the evolutionary benefits associated with this. Additionally, I am investigating how tourism and other human activities impact the tree-climbing habits of lions in the region. By unravelling these connections, we can gather valuable insight that informs conservation efforts and promotes sustainable practices in Ishasha and other ecotourism destinations.

Lions are often covered in flies, especially after feeding. This may be one of the many reasons why they climb trees. © Ross Couper

DEMOCRATIC REPUBLIC OF THE CONGO

18

Lions of hope in the Virungas

The Lions in the Virunga National Park, in the Democratic Republic of the Congo (DRC), are separated by a small, easily crossable river from the lions of Uganda's abutting Queen Elizabeth National Park. However, it's not easy to be a lion in a war zone, even when peace breaks out. Same guns, different people.

Olivier Mukisya, Virunga National Park

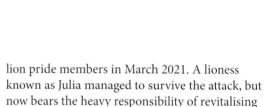

Virunga National Park – a UNESCO World Heritage Site and one of Africa's most biologically diverse protected areas – has been deeply impacted by the effects of war and armed conflict for over 20 years. It is now protected by a dedicated team of over 700 rangers.

These men and women, all of whom are Congolese and drawn from the areas surrounding the park, go through intensive training, risking their lives daily to safeguard the exceptional wildlife, flora and fauna, including some of the last of the world's endangered mountain gorillas.

Throughout the storied history of the Ishasha Valley (just across the border from Uganda's Queen Elizabeth National Park), lions have played a leading ecological role despite fluctuations in population size. Natural threats to their survival continually test their resilience in this wild savanna habitat, with only a fraction of cubs reaching adulthood.

Prides here must also contend with obstacles related to human activities, including the formidable problem of illegal poaching. Tragically, one incident alone claimed the lives of six young lion pride members in March 2021. A lioness known as Julia managed to survive the attack, but now bears the heavy responsibility of revitalising her pride.

So far, despite having given birth to multiple litters, none of her cubs have survived beyond infancy. Not only must Julia endure threats from humans, other animals and the harsh environment, but also threats presented by her own species: male lions periodically invade the pride's territory, injuring and sometimes killing her cubs, or those of other matriarchal groups.

In August 2023, Julia gave birth to four cubs, including Hope. Her survival alongside her siblings was precarious from the outset; tragically one of her brothers met his end at the horns of a buffalo. Still Hope persists, both physically and metaphorically, taking on allo-parental responsibilities, providing assistive care for her mother's most recent litter.

At under four years old, Hope is a torchbearer for the Ishasha lion population. With each passing day, there is cautious optimism that she may one day contribute to the restoration of lion numbers across these vast savannas of eastern DRC.

Just across the border from Queen Elizabeth National Park is the Virunga National Park in the Democratic Republic of the Congo, where lions are under threat.

Lions of hope in the Virungas

PORTFOLIO

It's not only leopards that hang out in trees

For the past 50 years, the Ishasha sector of Uganda's Queen Elizabeth National Park has been renowned for its iconic tree-climbing lions. The lions here climb large fig trees to rest, to escape the heat and flies and to scan the plains for their next prey. However, while Ishasha is the best known park for tree climbing lions, it doesn't have a monopoly as they can also be found in many other parts of Africa as these photos from Tanzania demonstrate. Nor is their behaviour restricted to trees. Any boulder, kopje, guest tent deck and even heavy earth moving equipment that can give them the height advantage or get them away from the flies will do.

It was thought that Queen Elizabeth National Park had the monopoly on tree-climbing lions, but that myth was dispelled long ago. This "Christmas Tree of lions" was photographed in the Serengeti.

Ndutu area, Serengeti National Park, Tanzania. © Marlon du Toit

Serengeti National Park, Tanzania. © Ariadne van Zandbergen

Ngorongoro Crater, Tanzania. © Bobby-Jo Vial

Magashi, Akagera National Park, Rwanda.
© Marcus Westberg

Akagera National Park, Rwanda. © African Parks

19

Lions of Akagera

The Rwandan genocide had a devastating impact on the country's wildlife. Lions have been central to bringing its most important national park back to health.

Drew Bantlin, Rob Reid & Andrea Reid

Founded in 1934, Akagera National Park once spanned nearly 10% of Rwanda, encompassing a mosaic of bushland, open grassland and diverse forest habitats. Lion populations thrived there and, as recently as the early 1990s, were estimated at 300 individuals.

The 1994 genocide left Rwanda in a state of total devastation. With millions of internally and externally displaced people, the decision was made to allocate two-thirds of Akagera as resettlement land. As refugees returned to the area, so did their cattle. More than 30,000 head were estimated to be grazing inside the shrunken boundaries of the park. Human encroachment into the remaining habitat and predation on livestock by lions led to widespread retaliatory killing, resulting in the species being extirpated from Rwanda by 2001.

The following decade saw wildlife populations in Rwanda decline by an estimated 80% across all species. Black rhinos were poached out of the park in 2007 and all large mammals declined in the face of unchecked poaching, grazing competition and habitat degradation. Then in 2010, African Parks assumed management of Akagera in partnership with the Rwanda Development Board.

Akagera is Rwanda's last protected savanna; securing the park was the most urgent mandate to the new management. Upscaling of ranger teams and law enforcement capacity addressed the initial challenge of poaching in the park. In tandem with widespread community initiatives, trust was slowly built between management and the communities. With a focus on empowering communities through conservation, using a three-pronged approach of enterprise, engagement and education, support for the park has been achieved across the communities living adjacent to Akagera.

Ahead of lion reintroduction, widespread sensitisation programmes were carried out in the communities to understand community members' perceptions and concerns about lions, to generate

Akagera National Park, Rwanda. © Marcus Westberg

awareness for the translocation and how to mitigate potential human–lion conflict. Local meetings, presentations and film shows were held to increase understanding of the role of lions in the ecosystem.

The potential for human–lion conflict was discussed and how the park and communities would work together to mitigate it. A 120km predator-proof fence was erected along the western boundary of the park and signage encouraged community members to keep their livestock away from the boundary. Support for the lion project was gained and pride in their reintroduction grew.

With this support, poaching greatly reduced and management capacity elevated, seven lions were translocated to the park from South Africa in 2015. They quickly settled into Akagera. Initially, the five females ventured widely, nearly reaching the southern boundary some 50km from the release site, while the two males moved north. Within a month, the wandering period began to wane and the lions settled into a core home range in the north of the park close to areas of high prey density, good water availability and potential den sites. The males re-joined the female group and Akagera's first lion pride in nearly two decades was realised.

A natural fission-fusion social dynamic developed as the females dispersed and returned to the group within the larger home range. This exploration and realisation that they were the only lions in the system eventually led to three distinct female groups forming. The dominant female in the group, Shema, moved to the northern extreme of the park and settled into an area of good prey numbers and denning sites.

The sisters Kazi and Umwari branched off, moving south into the rolling hills of the park. And the two youngest females, Amahoro and Garuka, settled into a core range around the release site. The two males, Ntwari and Ngangari, floated between the females, ranging widely and associating with any female in oestrous. Their efforts bore fruit and less than a year after the reintroduction, Shema gave birth to three cubs. Kazi, Umwari and Amahoro all produced cubs in 2016 as well, and the population grew rapidly.

Aside from reproduction as the obvious indicator of success, close monitoring of the lion population showed normal behaviour, sociality,

> **Lions are apex predators and their reintroduction potentially has wide-reaching consequences for Akagera's environment.**

territoriality and hunting. All initial signs of the translocation were going well.

In time, two main prides consolidated, Kazi and Umwari in the south and Amahoro's pride in the north. There is a clear boundary to their ranges following landscape features and both prides are reluctant to venture too far into the other's claim. This spatial separation continues to the current day and all three female lions continue to guide their respective prides.

As the two males Ntwari and Ngangari have aged, their pride dominance has slipped away. A formidable young coalition of three males, sired by Ntwari and born to Kazi, have assumed control of the north of the park, pushing out their father's coalition and joining Amahoro's pride. In the south, Kazi and Umwari's group was joined by a coalition of two males, two of the first cubs born to Shema in 2016. This pair, also sired by Ntwari, have pushed the original males out and continue to patrol the bounds of their territory with ferocity.

Beaten and ageing, Ntwari and Ngangari continue to live in refuge, moving to avoid the younger, stronger males and subsisting primarily on hippo carcasses they scavenge on Akagera's remote peninsulas. Now over 12 years old, their days are likely numbered.

Numerous litters of cubs have been born and cub mortality is low. 2019 saw the first 'cubs born to cubs'; the start of the second generation of lions in Akagera. Social structure has remained consistent and normal as the prides have grown. Second-generation males have started dispersing and two young coalitions of males now use the middle of the park. While they may currently be dodging the more dominant coalitions, they have started ranging further into the prides' territories. It won't be long before the next pride-turnover event, when one of the younger coalitions will challenge the incumbent coalitions for access to mates and control of the prides.

The natural processes of cub dispersal, pride turnover and inbreeding avoidance are functioning in the population. However, in a small, fenced reserve these processes can falter or become non-existent. Future management actions will focus on assisting these processes and may include additional translocations as a management tool. In a secure and prey-dense park like Akagera, loss of these important processes stands to be a potentially important threat to population persistence.

Diet has also developed in Akagera's lions, as expected. Akagera harbours a large, diverse population of herbivores and, in the early days following the reintroduction, the lions preyed on an array of species. Influenced both by the lions' exploration of the landscape and the prey's naivety towards these big cats, monitoring teams observed the lions feeding on everything from impala and topi to waterbuck, zebra and buffalo. In time, their diet stabilised and, as expected from studies on diet in well-established systems, the lions began showing a clear preference for certain prey species. Warthog, zebra and buffalo now make up nearly 80% of the lions' diet, while they avoid the smaller, lighter species, such as impala, which are simply too fast for the lions to catch.

The diets of specific prides also evolved. Pride 2 has notably shifted their diet over the years as pride composition has changed. Two years ago, Pride 2 preyed mainly on warthogs, which comprised more than 60% of their meals. At a time when the pride consisted of three adult females and 10 small cubs, the females struggled to hunt larger species. Instead, needing to provide for many mouths, they developed methods for digging warthogs from their burrows. As the cubs grew, this necessitated a shift towards larger prey that could provide for a growing pride. Today, the diet of Pride 2 has flipped to being more than 60% buffalo. Variation further exists between the prides; Pride 1 consumes zebra as nearly 50% of their diet, with buffalo and warthog being minor, but important, secondary options.

As the lion population continues to grow and diets shift with changing social compositions, the continued gathering of data to inform management decisions will become even more important. Routine close monitoring has contributed to this deep understanding of the lion population in Akagera, and the teams on the ground tracking the lions play a vital role in ensuring that the principles of adaptive management can be applied in the decision-making process.

Managing the lion population will become more challenging and necessarily more hands-on as the population increases, given the relatively small size of Akagera. While initial management focused on ensuring the persistence of lions in the landscape, this focus will shift towards a holistic management strategy, with lions viewed as a functioning component of Akagera's ecosystem.

Lions are apex predators and their reintroduction potentially has wide-reaching consequences for Akagera's environment. This includes direct effects on herbivore populations and sympatric carnivores and possible knock-on effects – often positive – for vegetation and ecosystem-level processes. Understanding these effects will be key in evaluating the long-term success of the reintroduction and our understanding of how lions influence the landscape.

Akagera stands to be an important case study for informing other long-range translocation projects and re-establishment efforts of large carnivores into Central Africa. Continued evaluation of the success of lions in Akagera will be essential as they reclaim their role at the top of the trophic pyramid. Until this can be fully evaluated, the short-term achievements of no human–lion conflict, no lions lost to poachers and numerous litters of cubs born will continue to be motivation for the monitoring teams and park management.

All three of these achievements are testament to the efforts of park management, the pride of Akagera's communities in having lions in the park and wide-reaching support for conservation in Rwanda, making this country one of the global leaders in the field.

Akagera National Park, Rwanda. © Adriaan Mulder

Serengeti National Park, Tanzania.
© Johan van Zyl

The lions of Tanzania

PORTFOLIO

Wildebeest emerge onto the plains in the Ndutu area, southern Serengeti National Park, Tanzania, as one of their number is attacked by a lion on the hill. © Michael Poliza

Selous Game Reserve, Tanzania.
© Robert J Ross

Southern Serengeti National Park, Tanzania. © James Lewin

Serengeti National Park, Tanzania.
© Michael Poliza

Queen of Utafiti Kopje, Serengeti National Park, Tanzania.
© James Lewin

Nyerere National Park, Tanzania. © Robert J Ross

Southern Serengeti National Park, Tanzania. © Michael Poliza

TANZANIA

20

Living with lions in Ngorongoro

Creating a working relationship between lions and pastoralists takes time and patience.

Sally Capper, KopeLion

Alongside the Serengeti National Park, the Ngorongoro Conservation Area (NCA) – a UNESCO World Heritage Site – is unique among the world's savanna ecosystems. Covering some 8,300km², it is characterised by seasonal water availability. It is influenced by the migratory patterns of wildebeest and includes a renowned lion population.

Home to nearly 100,000 traditional pastoralists, mainly Maasai and Datooga, the NCA is a multi-use protected area. These communities rely on livestock for their livelihoods, creating a delicate balance with the ecosystem. While lions contribute significantly to tourism revenue, their attacks on livestock pose a substantial financial burden, leading to slaughter of the predators for self-defence and in retaliation.

Despite the area's natural and cultural uniqueness – acknowledged by its triple status as a UNESCO World Heritage Site, International Biosphere Reserve and Global Geopark – local communities feel marginalised in conservation decisions, lacking sufficient benefits amid prevailing conflict and poverty.

The NCA is a critical hub for the dispersal and connectivity of the Maasailand lion metapopulation spanning northern Tanzania and southern Kenya. However, human–environment dynamics, encompassing climate change, rangeland degradation, population growth and direct conflict have resulted in the loss of lions from much of their historical range within the area. The high risk of human–lion conflict, especially, affects dispersing lions, compromising the connectivity between the crater lions and the broader metapopulation.

Reducing lion killing

In response to these challenges, conservationist and National Geographic Explorer Ingela Jansson based herself in Ngorongoro in 2011 to continue long-term monitoring of the crater lions, expanding the project to include the NCA multi-use area and the local pastoralist community. In 2014, through collaboration with the Lion Guardians from Amboseli in Kenya and the adoption of their model of engaging pastoralist warriors in lion monitoring and supporting communities to coexist with lions, she founded the Korongoro People's Lion Initiative (KopeLion).

As part of the greater Serengeti ecosystem, the NCA is a crucial site for studying and establishing a sustainable coexistence model, ensuring meta-population connectivity. KopeLion aligns its efforts with long-standing culture and traditions, focusing on three key areas: cutting costs, realising value and applying knowledge.

While traditional lion hunting (*ala-mayo*) has largely ceased, retaliatory killings persist owing to livestock losses. KopeLion aims to reduce the loss of livestock to lions and therefore prevent these killings. At the core of this work are the *Ilchokuti* (guardians of the livestock and lions), who continuously monitor the presence of lions, including GPS-collared lions, and communicate their whereabouts to assist herders in protecting their livestock.

They also offer support in repairing livestock enclosures breached by predators, treat wounded livestock and find livestock that is lost and vulnerable to attack. The initiative engages communities in lion conservation, emphasising the recognition of lions' worth.

There are costs to existing alongside large carnivores and the benefits are sometimes less obvious. KopeLion aims to boost both tangible and intangible benefits (such as employment providing ecosystem services) that will ensure lions are seen as a gain for pastoralist communities.

KopeLion, in collaboration with the TAWIRI (Tanzania Wildlife Research Institute) lion research project and the NCA Authority, integrates scientific and traditional knowledge in designing, implementing, measuring and applying coexistence strategies. The lion population in Ngorongoro Crater has been studied continuously since 1963, with the Ndutu lions monitored since 2010.

The focus is on recognising each lion individually and tracking them throughout their life for detailed data on population trends and status. This approach provides opportunities for measuring the effects of coexistence strategies across multi-use areas, ensuring connectivity across large landscapes and ultimately serving the lion population.

By addressing the intricate web of challenges facing this region, KopeLion attempts to ensure a viable coexistence of pastoralist communities and lions, preserving the unique ecological balance of the Ngorongoro Conservation Area.

KopeLion monitors keeping track of lion movements, Ngorongoro Conservation Area, Tanzania. © Sally Capper

The NGO KopeLion collaborates with Ngorongoro communities to promote lion connectivity and enhance pastoralist livelihoods in the greater Serengeti, integrating cultural, environmental and economic values for lasting coexistence. © Ingela Jansson

TANZANIA

21

Traditional use of lion parts

The Sukuma are agro-pastoralists originating from the Shinyanga and Tabora regions of northwestern Tanzania, and are the largest ethnic group in Tanzania.

Jonathan L Kwiyega & Belinda J Mligo

In the 1960s and 70s the Sukuma population grew rapidly in numbers and they now constitute between 13% and 20% of Tanzania's population. Their gradual spread across Tanzania has pushed back margins of undeveloped land, attracting considerable concern over their environmental impact.[1]

They are renowned for their cultural dances, a practice ordinarily engaged in by young men and associated with the use of traditional medicines and animal body parts, deemed to enhance their strength, agility and ability. These dances fulfil several roles, ranging from the celebration of a good day's farming, or advertisement of feats of bravery, to annual competitive dances. Dancing is frequently associated with the payment of a reward – often cattle – and in this sense, it becomes linked to lions.

While some tribes like the Kaguru believe in the existence of spirit lions, which protect them against the possible attacks by normal lions and protect crops from wild pigs, the Gogo tribe in Dodoma region use lion fats to give them courage.

In western Tanzania, local traditions and culture permit both non-retaliatory and retaliatory lion killings. Retaliatory killings make up the majority of anthropogenic lion mortalities and have a marked negative impact on population numbers.[2] With habitat loss and rapid population growth of both humans and livestock, threats from and to lions are escalating.[3] To kill a lion (known as *mwizi wa ng'ombe* or 'cow thief') in defence of your or your neighbours' cattle is seen as the ultimate act of bravery.

Lions in decline

Around 37% of Tanzania has been set aside for wildlife and biodiversity conservation. This includes national parks, game reserves, game control areas, wildlife management areas or open areas.[4] The country has the largest population of wild lions in Africa, equal to about half the global population.[5] However, being apex predators, lions and other carnivores are more threatened than other mammals because of growing human and livestock population demands, and this is leading to escalating human–wildlife conflict (HWC).[6]

In western Tanzania, predators are threatened by habitat loss, a declining prey base, increased

196 ~ THE LAST LIONS

interactions with humans and livestock, legal hunting by tourists and illegal traditional hunting.[7] In addition, culture and traditions, local beliefs and bushmeat preferences among community members living alongside reserves are major factors in their decline.[8]

Sukuma agro-pastoral communities, which are growing around protected areas, keep their livestock in marginal habitats and commonly adjacent to protected areas. Pastoralists also believe that grazing in reserves improves the breeding potential and productivity of their cattle. All this makes their livestock highly vulnerable to predation by lions.

Recent manipulation of Sukuma traditions regarding lion hunting has negatively impacted wildlife conservation and, as pastoralists, the Sukuma understandably harbour negative attitudes towards livestock predators.[9] There is little independent lion population monitoring and limited resources and information for conservation interventions, so lions will continue to face high threats in the landscape. With declining prey species and increased settlements, lions are being hit by revenge killing and removed as enemies of both humans and livestock. When herders lead their cattle into nearby reserves in search of pastures, risks to lions escalate as they shift their diet to easily caught livestock. There are three distinct motives for killing lions in Tanzania:

- Retaliation for attacks on livestock[10] or humans.[11]
- Tourist trophy hunting, long thought to be of negligible significance, is drawing renewed attention as a significant threat in some areas.[12]
- Traditional hunts, often linked to rites of passage, on which this chapter will focus.[13]

Lion's mane headdress. © Brent Stirton

Traditional use of lion parts

Among pastoralists, trophy-taking of lion parts signals honour and prestige, making the killer a hero.[14] Such killings gave rise to one of the most famous Sukuma dancing rituals, the Lion Dance. Dancing by the lion killer serves multiple purposes. First, it advertises his bravery. It is thought to stave off the impending madness that would otherwise occur if the spirit of the lion possessed its killer. In reward for his bravery, a lion killer's relatives are obliged to bestow gifts of cattle as a reward. Therefore, killing a lion almost guarantees the accumulation of wealth with which brides can be bought, fortunes made, and fame assured.[15]

A lion dancer will be required to provide evidence of the killing. Among these are the symmetrical skin from nose to tail tip, lion fat, heart, teeth, claws, hair and tail, used directly or as amulets. He is smeared in lion oil, will not wash or engage in sex between the killing and the dance, and will dress to reflect a lion-like appearance.

There is a strong belief that it's possible to absorb harmful elements from the environment that can cause misfortune and ill health. However, by taking a certain medicine, it's considered possible to guard against malice and bring good luck and fortune. Such protection can extend to guarding against witchcraft, evil spirits, jealousy and infidelity and strengthen love affinity and job opportunities.

Commercial, trade-driven poaching for body parts is therefore a major cause of population decline for several big cats in western Tanzania.[16] Many lion body parts are used in traditional ceremonies, rituals and as decorations, encouraging black-market trade.[17] Teeth and claws tend to be internationally traded – generally to Asian countries and particularly Vietnam – which is driving poaching.[18]

In Africa, the use of lion derivatives is widespread. A study in Nigeria in 2009, found lion fat being used to treat back and joint pain; the skin and lungs to treat whooping cough; lion veins to treat erectile dysfunction; lion noses to treat stomach problems; and lion livers to treat headaches.[19] Lion fat use is prevalent in South Africa among the Xhosa, and it is used as protection against evil spirits in Zimbabwe. The Khoisan spread it on their body to detect if a lion is near while walking in the bush. The Herero in Namibia and Botswana also use lion fat and, in Kenya, the Samburu use it to keep away the creditors as the scent of the fat is said to evoke fear.

Attempts to halt the use of lion parts

Tanzania has by-laws that seek to protect lions from traditional lion hunting and killings. These laws ban practices that encourage and support the traditional use of body parts as well as lion dancing and rewarding of lion hunters or offtakes from trophy hunts.

In step with these laws, village-level conservation groups participate in lion monitoring and inform farmers of the presence of lions. Where there is no cellphone coverage, loudspeaker systems have been installed to transmit lion-related information through public announcements.

These groups engage in sensitisation and education, teaching communities about conflict mitigation, lion behaviour, risky human behaviour and seasonal movements of lions. In western Tanzania, this type of information is included in the curriculum in schools and spread through park visits, film shows and by sharing educational materials such as booklets and brochures. This has improved community perception of the value of lions and has resulted in a decline in negative interactions.

The first of such programmes was introduced by the WASIMA campaign (Watu, Simba na Mazingira – People, Lions and Environment), which focused on establishing protection by-laws in Mpimbwe, south of Katavi National Park. Village game scouts have been trained as Lion Conservation Ambassadors. Altogether, the campaign has successfully reduced traditional lion killings in remote places around protected areas there.

Recommendations

Lion killings and the use of body parts are largely an education issue, coupled with a lack of alternative income sources and livestock pasture. These factors feed traditions and customs that are inimical to lion conservation. To address retaliatory killings, we need to install more lion conservation and monitoring programmes in areas such as the greater Mikumi-Selous ecosystem and the existing Rukwa-Katavi-Mahale-Ugalla-Kigosi ecosystem, where locals still use lion body parts. All African countries with lions need to develop national lion conservation plans that guide citizen actions and scientific research studies as well as conservation and management.

Hunting regulations must forbid the hunting of lions younger than six years, hunt operators must be charged fees per lion hunt, and annual fees – regardless of offtake and trophy-hunting fees – must be greatly increased. We have to ensure effective ecosystem protection, management and conservation, capable of maintaining the food chains where lions have access to their preferred prey species and space.

Local communities should be helped to participate in environmental and wildlife conservation activities such as beekeeping, patrols, forest or natural habitat preservation, tree planting, alternative livelihoods, community conservation banks, incentives and credits that reduce pressure on nature. People need to be compensated for losses to lions. More generally, they need better health facilities and school buildings as well as assistance in drilling wells and farming fish and poultry.

With regard to resolving lion conservation issues, multi-stakeholder involvement is essential to ensure informed and effective implementation, with sensitivity to diverse cultural backgrounds. Land-use plans need to be established and adhered to, and local political leaders should discourage community members from encroaching on, or settling in, wildlife corridors, reserves and dispersal areas. Land use planning should include regulation of immigrants from other areas to prevent movement into vacant land abutting protected areas.

Strengthening wildlife laws should include enforcement of border control and customs security at all ports (sea, land, air) to improve detection and control of illegal wildlife trade and to gather information on trafficking routes and commodities. To this end, governments and NGOs must allocate resources towards staff training in detection, screening techniques and other such technologies. Local, national, transboundary and Africa-wide collaboration has to be improved to ensure seizure data is accurately stored and accessible.

What is critical to any success, however, is the engagement of local communities at all levels.

Maasai warriors wear traditional lion's mane headdresses during important ceremonies and events. These headdresses are a symbol of strength and bravery reflecting the warriors' connection to the powerful animal in their culture.

Traditional use of lion parts ~ **199**

Rite of passage

Saning'o Kimani, KopeLion

At the end of June 2024, a historic event unfolded within the Ngorongoro Crater when over 1,000 young men, aged 15–24, gathered to participate in the culturally significant rite of passage known as *Engipaata*. This occurs every 16–20 years and is monumental for the Maasai culture as it prepares the youth to become the next generation of warriors. I was born and brought up in the Ngorongoro Conservation Area and am KopeLion's programmes co-ordinator. KopeLion was invited to the event and was welcomed by the *Alaigwanani* (the chief) to help promote peaceful coexistence between the community and lions by showcasing films highlighting this message. The *Alaigwanani* also took the opportunity to address the young boys, emphasising that the era of demonstrating bravery through lion hunting has passed. He underscored that true bravery in the modern age involves staying in school and pursuing education.

Young warriors gathering in Ngorongoro Crater, Tanzania, to participate in an important rite of passage ceremony.

In Tanzania's Ngorongoro Crater, young Maasai are addressed by their chief (*Alaigwanani*) during a rite of passage ceremony that will prepare the youths to become the next generation of warriors. © Saning'o Kimani, KopeLion

TANZANIA

22

Dam and be damned: Stiegler's Gorge

They were two presidents of Tanzania whose goals could not have been more different: one supported the protection of the greatest game reserve in the world, the other sliced it in half, dammed the river and sold the timber.

Colin Bell

In his 1961 Arusha Declaration, the president of newly independent Tanzania, Julius Nyerere, laid out his country's conservation position: 'The survival of our wildlife is a matter of grave concern for all of us in Africa. These wild creatures amid the wild places they inhabit are not only important as a source of wonder and inspiration, but they are an integral part of our natural resources and Tanzania's future livelihood and wellbeing. In accepting the trusteeship of our wildlife, the government of Tanzania solemnly declares that we will do everything in our power to make sure that our children's grandchildren will be able to enjoy this rich and precious inheritance … the success or failure of which not only affects the continent of Africa, but the rest of the world as well.' [1]

How ironic that Tanzania's destructive new hydroelectric dam, which is being constructed on the Rufiji River in the heart of the Selous Game Reserve, is named the Julius Nyerere Hydropower Project after the ex-president who fully understood the value of preserving Tanzania's wild places long before conserving biodiversity was recognised as vital for the health of the planet and its people.

Since its inception in 1896, the Selous Game Reserve in southern Tanzania has been one of the world's largest and most important wildernesses and a key protected area for both wildlife in general and lions in particular. Between 1951 and 1973, the park went through a period of rapid expansion and was enlarged to 50,000km^2 to become a massive contiguous tract of unspoilt, protected wilderness. For the next 50 years, the Selous was larger than nearly a third of the countries in the world. In 1982, it was declared a World Heritage Site by both the Tanzanian government and UNESCO.[2]

Lions in the Selous

In 2006, International Union for Conservation of Nature (IUCN's Species Survival Commission Cat Specialist Group identified the Selous Reserve as a key lion conservation unit; and, because it was the largest and most contiguous protected area in Africa, it was arguably the most important sustainable lion range in Africa. The majority of lion experts agreed the Selous contained the greatest number of wild lions in Africa, but because of

Selous Game Reserve, Tanzania.
© Dana Allen

Against all environmental and economic advice, the construction of Stiegler's Dam went ahead in the very heart of Selous Game Reserve, Tanzania, destroying over 300,000 acres of prime wilderness.

The construction that has taken place in the heart of the old Selous Game Reserve in Tanzania is one of the biggest environmental travesties of this generation.

the reserve's sheer vastness, its rugged terrain and inaccessibility, no one knew how many. Lion experts estimated the reserve contained anywhere between 3,000 and 4,300 individuals.

Hans Bauer, an experienced lion researcher, calculated in 2016 that, between 1993 and 2014, lion numbers throughout Tanzania had declined by 66%, largely through people pressure and habitat loss, making the Selous much more important for lions in Tanzania and, indeed, Africa.

How do we accurately count lions in such an enormous, rugged and often inaccessible chunk of some of the wildest parts of Africa? A study conducted in 2018 in an effort to estimate the reserve's lion densities required extrapolation of numbers from a selection of smaller sample areas covering about a quarter of the Selous. Results suggested that the number of lions in the sample areas canvassed varied from 2.1 to 5.1 individuals per 100km². Averaging this out over the entire reserve would indicate that there were around 1,800 lions throughout the reserve.

All population estimates need to be treated with a certain degree of scepticism. This particularly applies to numbers gathered and then extrapolated from the 2018 study, 'because our sampling intensity was not distributed evenly over the study area'. Whatever the numbers, in terms of lion conservation Selous has always been a prime lion stronghold.

So, what has the construction of the dam got to do with lions and their dynamics? Stiegler's Dam, now completed, has created an impenetrable 90km lion-proof deep-water barrier, severing Africa's largest and most important lion refuge into two distinct parts. In parts, the dam is up to 30km wide, shutting down many prime lion habitats, ancient lion and elephant migration routes and gene pool exchanges.

How did the Tanzanian government and the world allow this to happen? Trouble started brewing in 2010 with threats of a plan to dam the Rufiji River at Stiegler's Gorge to create hydroelectricity. If the plans went ahead, the dam would be the ninth largest globally when completed. However, many of the Tanzanians and Selous safari operators we canvassed felt the dam would not happen because the country could not afford the billions of US dollars needed to fund its construction.[4]

The viability and practicality were also questionable because of the distance from the gorge to where the electricity was needed in faraway industrial hubs like Dar es Salaam. And anyway, UNESCO World Heritage status and protocols prohibited any such development. Surely no rational politician or government would go ahead with such a costly and destructive project against the advice of just about every economist, environmentalist, climatologist and even against the advice of the United Nations (UN) and World Bank's best scientists and experts.

But everyone underestimated the drive of newly elected President John Pombe Magufuli, who was voted into power in 2015 on a developmental ticket that favoured large-scale state-controlled industrialisation and developments. Magufuli was a former minister for roads and infrastructure, renowned for his lack of regard for nature, its intrinsic values or its ecological and economic importance. He was a nationalist to the core and when quizzed about the negative environmental impacts his projects would have on biodiversity and conservation, he said, dismissively: 'Nobody can teach us about conservation'.[5]

A year after his inauguration as president, the construction of the Selous dam started. Tender documents for its construction were announced in 2017. I wrote about the damages the dam would cause in the book *The Last Elephants* and lobbied non-government organisations (NGOs) and politicians to see what could be done to stave off such a destructive development.[6] Resource economist Dr Ross Harvey wrote a well-reasoned paper about why the dam should not be built, noting there were much more effective and efficient ways to create electricity for Tanzania through solar, wind, smaller hydro dams and by utilising the country's abundant natural gas reserves.[7] Harvey argued that ample 'green' electricity could be produced closer to where it was needed at a fraction of the cost of Stiegler's Dam and with almost zero environmental damage.

However, no one factored in the determination of Magufuli, aptly nicknamed The Bulldozer. He made Stiegler one of his government's flagship developmental projects. Nothing got in his way once he had made up his mind – not even COVID-19. As one of Africa's prominent coronavirus deniers, he declared early in the pandemic that 'the corona disease has been eliminated' throughout Tanzania and that the country was COVID free.[8] Ironically, just eight months later, COVID-19 could have been the cause of his death, even though the official cause was put down to heart complications. Few Tanzanians believed it was true – some even suspected he was poisoned.

Once the realisation that Magufuli was going ahead to approve the building of the dam and 121,406ha would be denuded and three million trees would be chopped down to make way for it, opposition started building both within Tanzania and globally. In June 2018, the IUCN publicly declared that 'Deforestation of 140,000ha in the core of the Selous Game Reserve will clearly put the Selous ecosystem and its wildlife in great danger. This is one of the planet's most biodiversity-rich areas and wise decisions are needed to ensure its protection for future generations. The logging is planned in the heart of Selous Game Reserve along the Rufiji River in an area containing important habitat for iconic mammals.'[9]

In reply to previous concerns about losing UNESCO World Heritage status, Natural Resources and Tourism Minister Ezekiel Maige brushed off the UN's World Heritage Committee as an 'insignificant entity from which we cannot take orders'.[10] In mid-2018, environmental minister Kangi Lugola warned that anyone who resisted the project would be jailed.[11]

The government under Magufuli charged ahead despite a deeply flawed Environmental Impact Assessment process and cost estimates calculated by independent experts to be more than US$6 billion more than the government's rather implausible US$2.9 billion budget. Given the state of the country's economy, its limited tax base and the World Bank's refusal to fund the dam, there are questions about whether the government has sufficient money to fund the project in its entirety or to repay any large loans.

To keep international NGOs quiet and onside, the Tanzanian government enacted a shrewd manoeuvre in 2019 by renaming, recategorising and gazetting the western sector of the original Selous Reserve to become the new Nyerere National Park. Being a national park under the management of the Tanzania National Parks Authority (TANAPA) theoretically gives the new national park more protection than what was previously possible without national park status. With local NGOs effectively muzzled, this tactic was enough to keep the major international NGOs from mounting a determined international PR campaign to try and stop the construction of the dam. The now much smaller Selous Game Reserve to the east of the dam remains a game reserve in status under the management of TAWA – Tanzania Wildlife Management Authority – which, in theory, has less legislative protection than national parks. Time will tell if the new Nyerere National Park and the 'new', much smaller Selous Game Reserve will survive future pressures and remain secure wildlife and biodiversity sanctuaries where lions, in particular, can thrive.

At the time of writing in June 2024, the Stiegler's Gorge Dam is nearing completion despite running well behind schedule. The dam, at 130m high and 700m wide, has reached full capacity a lot quicker than most predictions thanks to some tremendous local rains and the turbines are now being tested. The full dam has, as predicted, destroyed large parts of the world's pristine wilderness and carved up one of Africa's important lion sanctuaries.

There is no question that the Selous/Nyerere system, and all its lions, elephants and other wildlife, as well as the pristine wilderness, is at a crossroads. Even before construction of the dam began, a rich uranium deposit saw the boundaries of Selous shrink by over 24,281ha to allow for a new mine to be built. Along with the mining came permission to discharge poisonous and potentially radioactive effluent into the environment from a particularly toxic bleaching process that blasts water at the uranium-containing rock to separate it from the soils.

> **My guess is Tanzania's founding president Julius Nyerere would be turning in his grave if he knew what had just happened in Selous!**

In addition, nearly 50 mineral exploration licences have been issued in the region. Consider what will happen to these reserves when the next big mineral discovery is made. Before his death, Magufuli gave permission for a new commuter road to be constructed through the middle of the Nyerere National Park, connecting the towns of Ilonga and Liwale, located on opposite sides of the park. History elsewhere shows that when public access roads are built through parks, poachers, wood harvesters and others will make use of the fast, easy access to plunder.

There is also a plan to build the Kidunda water storage dam on the Ruvu River close to the northeastern boundary of Nyerere National Park to supply water for Dar es Salaam. Its headwaters could flood important dry-season grazing lands for wildlife within the Matambwe sector of the Nyerere National Park.

Beyond the threats to lion and wildlife populations, the destruction of three million carbon-sequestering hardwood trees has happened at a time when the planet is facing severe threats from carbon emissions, habitat loss and de-wilding. We are living in a time when every indigenous tree and every prime wilderness should be protected and celebrated not only for biodiversity, but also to help create more oxygen for the planet, while removing carbon from the atmosphere for storing in pristine African soils to help mitigate the threat of climate change.

Kenyan environmentalist and Nobel Prize winner Wangari Maathai's prophesy was on point when she wrote, 'The generation that destroys the environment is not the generation that pays the price'.

My guess is Tanzania's founding president Julius Nyerere would be turning in his grave if he knew what had just happened in Selous!

TANZANIA

23

Tanzania: and now for the good news!

What it takes to resurrect decimated wildlife populations from the ashes and then maintain them.

Chris & Monique Fallows

Most of the famous national parks and game reserves in Africa such as Amboseli, the Serengeti, Maasai Mara, Okavango Delta, Chobe, Etosha, Mana Pools, Hwange, Luangwa, Kruger and the Kgalagadi have established infrastructure and road networks, and are inhabited by animals that are used to the presence of vehicles or even people on foot.

Without giving it much thought, safari goers today think it normal that we can be very close to lions and all forms of wildlife that for the most part seem completely at ease with our presence. In these truly wonderful wildlife environments, it is easy to forget the hardy pioneers who stood tall to create parks and then protect them. We seldom have any clues as to the challenges and unbelievable difficulties they faced – and the hard work it took to make these prime wildlife refuges what they are today. It took the park managers of the old Selous Game Reserve 14 days to get to their main base camp, with the last five days on foot. Their patrols, all on foot, lasted up to three weeks at a time. In most of these parks, this immense struggle took place before we were born a half century ago. In some parks like Liuwa Plain in Zambia, the transformation is recent, but well underway. In other areas like Usangu and Msolwa, the journey has just begun.

Six Rivers Africa is a not-for-profit organisation, funded by Sir Jim Ratcliffe and managed by CEO Brandon Kemp and his team. In spite of the mountain of issues that need tackling, Six Rivers Africa is providing logistical, managerial and financial support for anti-poaching and research to Tanzania National Parks Authority (TANAPA) and Tanzania Wildlife Research Institute (TAWIRI) in both the Usangu and Msolwa areas. In Brandon's words: 'Our approach is to march forward hand in hand with the Tanzanian government in Usangu and Msolwa. With this approach we are building capacity in the Tanzanian organisations to

Ruaha National Park, southern Tanzania. © Ariadne van Zandbergen

better protect these vast wilderness areas for the future. Ours is a Tanzanian-led conservation and management approach.'

We flew towards Usangu and Msolwa in southern Tanzania, two of Africa's truest remaining giant wilderness areas – ones that most people have never even heard of. The Usangu Sector is in the south of the 24,000km² Ruaha National Park, while the Msolwa Sector is located in the remotest corners of Tanzania's largest national park, the giant 37,000km² Nyerere National Park, which evolved out of the old Selous Game Reserve. It is in these regions that only visionaries and the extremely brave put their shoulders to the conservation wheel.

You know you have landed somewhere remote and special when you are the only two people on a 14-seater Cessna Caravan aircraft and the seasoned bush pilot takes photographs of the area. I thought that, after more than 100 safaris to the great wildlife havens of Africa, there was little left to see or photograph that was unique. However, on our visit to Usangu and Msolwa, the penny dropped as to what truly wild places are really all about and why those of us who yearn for deep nature continue to strive for the tremendous rewards and natural highs we get from immersing ourselves in these wonderful wild places.

I had always thought of both Usangu and Msolwa as hunting areas that had been decimated by poaching, and where wildlife was so traumatised that it was, sadly, not worth the effort of going there. Our desire to visit was, rather, to see the incredible baobabs that Ruaha is famous for, and hopefully photograph wildlife in amongst these ancient giants. Arriving in Ruaha, you are instantly struck by the fact that there are not just a handful of these trees, but entire forests of them. We saw thousands of trees that had outlived dictators, wars, famine, feast, droughts and floods. To our amazement, in the riverine vegetation and forests lining the dry Mwagusi Riverbed we saw multiple lions, leopards and wild dogs, along with seldom-seen antelope like lesser kudu and Kirk's dik-dik. For anyone interested in photographing lion, leopard, elephant or buffalo with a multitude of other species thrown in, you won't be disappointed. The Usangu and the whole of Ruaha is certainly worth protecting.

Nyerere to the east of Ruaha is a national park that is more than 50% larger than Kruger National Park, and used to be home to Africa's greatest concentration of elephants – a population of over 120,000 individuals just a few decades ago. That population has dwindled to around 15,000 due to sustained poaching from the 1970s through to the early 2000s. Poaching also had the impact of rendering the black rhino extinct in this area.

Fortunately, the current Tanzanian government understands the potential of ecotourism and sees Tanzania's wildlife as one of its greatest remaining assets. Together with rangers and a dedicated team of veterinarians, coupled with the openmindedness of the government to work in partnership with foreign donors and teams like Sir Jim Ratcliffe's, significant programmes are afoot to tackle issues like logging, ivory and fish poaching, wildlife poaching and more.

The Msolwa area in western Nyerere is truly spectacular, with rivers that wind through the almost endless wildness. These are laden with iconic fish species such as tiger fish and vundu, and home to huge numbers of hippos, crocodiles and other wildlife. Lions are present, probably in great numbers, but are skittish and elusive as they have had little contact with humans and vehicles. Adjacent to the rivers are magnificent hardwood riverine forests, and inland from these the miombo woodland is punctuated by *bais* (waterholes) and grassland savannas. Msolwa is huge and diverse, begging to be nurtured and reborn.

The conservation challenges are significant, as was communicated to us by the TANAPA team. Huge swathes of land and primary forest adjacent to the park's boundaries are being cleared for cattle, sugarcane farming and for their hardwood, and the fingers of human impact reach illegally into the fringes of the park. Encouragingly, in the time that Six Rivers Africa has provided TANAPA with aerial and ground support, the number of poaching incidents has dropped significantly.

This firsthand insight into what could be the future of Nyerere was tempered by anxiety about what would happen if support from Six Rivers Africa dried up, and TANAPA did not patrol to keep human encroachment at bay. It made us realise just what was at stake and how important the Six Rivers project was to ensure

> In both Usangu and Msolwa we met true heroes fighting for the protection of the wild, its biodiversity and its apex predator – the lion. And, by virtue of all that, they fight for the future of humanity too.

a more sustainable future for one of the world's few remaining truly wild areas. Its either that or a once-off calamitous exploitation and irreparable rape of a wilderness.

The Six Rivers Africa team has various projects aimed at empowering communities living adjacent to both Usangu and Msolwa. They sponsored the construction and running of a new school on the boundary of the park that helps to educate future generations and inspire them to protect what makes southern Tanzania so special. This is backed up with guide, chef and tourism training initiatives providing for more sustainable livelihood options than poaching.

Research is vital to gaining a better understanding about what species are in an area, the interconnectedness between these species, and the myriad threats they face. Six Rivers Africa has also sponsored the collaring by TANAPA of lions and other iconic species in the area in order to get a better idea of these animals' movements and spatial utilisation of the wilderness, and how the team can go about mitigating against human–wildlife conflicts.

During our stay at the Six Rivers Kilombero Camp in Msolwa, we joined the team on an elephant-collaring exercise taking place in an area that had been turned into sugarcane plantations by the Kilombero Sugar Company, which is majority owned by the Illovo sugarcane company in South Africa. This plantation has effectively divided a wilderness area in two, creating the inevitable scenario of elephants pillaging the tasty crops as they traverse previously wild areas. Later that day we saw a group of five male lions, and joined the research team on several tiger fish-tagging trips that will enable conservationists to assess their migratory and seasonal movements.

Above all, we truly hope that other visionaries will come on board to complement the efforts of TANAPA, TAWIRI and Six Rivers Africa. Much of Africa's environment and wildlife is fragile, and we can only hope that future political regimes will have the same foresight to welcome collaboration and help, the fruits of which were so clearly being demonstrated within both the Msolwa and Usangu sectors.

Our once-in-a-lifetime visits to Msolwa and Usangu have enabled us to experience the rebirth of two of our planet's great wilderness areas.

The old Selous Game Reserve is a giant that has been crimped, kicked, punched, dammed, flooded, logged and set on fire, but for now it is healing and waking up through the care of dedicated foot soldiers restoring this Eden to its former glory. We witnessed the effort it takes to resurrect decimated wildlife populations from the ashes, and then maintain them. In both Usangu and Msolwa we met true heroes fighting for the protection of the wild, its biodiversity and its apex predator – the lion. And, by virtue of all that, they fight for the future of humanity too.

An aerial view of the Luengue-Luiana National Park, with the Kwando River in the distance. Luengue-Luiana is currently a mixed-use park, and is not only home to a variety of wildlife but also has a number of small settlements of humans and their livestock. Although wildlife numbers in the area have been on the increase since the end of the Angolan Civil War, lion populations remain very low and sightings are rare.

24

Where are Angola's lions?

There were assumed to be over 1,000 lions in Angola, but when trained researchers hit the ground, quite a different story was revealed.

Dr Paul Funston

Thirty years of lion research and conservation has taught me that when it comes to lion population estimates, we tend to live in a dream world, often misled by 'expert opinion' expressed in maps and conservation strategies, and even in scientific publications.

In 2005 and 2006 scientists, country experts and government officials met in two momentous workshops: one in Johannesburg, South Africa, and the other in Douala, Cameroon. Their aim was to plot a strategy to save lions across the African continent.[1] In pre-workshop technical sessions, the participants drew on supportive information to compile maps showing everything that was known about lions at the time, to tally up the areas where lions still occurred and to provide estimates for how many still existed at each site.

This process built on the first attempts to develop a database of known lion populations in 2002[2] and continues to this day. The current database forms the backbone of vital estimations of how well lions are doing in Africa via the IUCN Red List process.[3] The database also directs conservation attention and helps shape requests for funding needed to tackle the recovery of some 60–70 remaining lion populations in Africa.

Although participants at these workshops brought knowledge and expertise to the table, there was much that was unknown, and guesswork filled in the gaps. This has caused some ongoing confusion, resulting in a mix on the resulting maps of detailed verified information along with some extrapolations of supposed lion range that have yet to be substantiated.

Angola's lions

One area that struck me was a vast swathe of central and southern Angola shaded in as potential lion habitat and supporting a possible 2,000 lions.[4] Had the civil war been positive for lions, or had it done what it normally does for wildlife and wiped it out? I suspected the latter and protested with co-authors in 2013 when a large area of Angola was shaded in to denote a possible contingent of more than 1,000 lions.[5]

That many lions would be Holy Grail stuff for lion conservation, with only the Okavango ecosystem, Kruger National Park, Selous Game Reserve, Ruaha National Park and Game Reserve, and the Serengeti-Mara ecosystem being in that esteemed league. The paper was published in 2006 and again in 2013, with Angola seen as a huge and significant country for lion conservation. But no one knew for sure, and nobody was working there to verify or refute these claims.

To the delight of conservationists worldwide, the Kavango-Zambezi Transfrontier Conservation Area (KAZA) was signed into reality in August 2011. Suddenly two 'partial reserves' and seven additional *coutadas* (hunting blocks) were lumped into two huge national parks (about 42,000km^2 each): Luengue-Luiana and Mavinga. I was living in Namibia's Caprivi Strip from 2013 to 2019 and worked for an NGO directing the strategy and implementation of its lion programme. I needed to venture into Angola in search of its lions.

I wanted to know if there were hundreds of lions there waiting to be discovered. According to the maps and other documents, 800 lions occurred just north of the border in the KAZA region adjacent to Namibia. With only about 60 heavily persecuted lions left in the Caprivi region, my suspicions were aroused.

Counting lions

In July 2013 I set off to explore Angola and get permission to count the lions there. After two years of pleading, frustrations, anticipation and disappointment, as well as experiencing the delights, extravagances and charisma of General Higino Carneiro, the governor of the Cuando Cubango region, I was finally invited to a signing ceremony at Dirico. Crossing the border from Namibia to Angola at this spot was a challenge, but when General Higino sends a small boat across the Kavango River and tells you to get into it and come across the border, well, you do it.

On the day of the ceremony, with much fanfare, pomp and ceremony, six military helicopters arrived bearing the elite, the press and members of the army and police. The signing ceremony was even attended by the president's son, José dos Santos. The key to unlocking Angola's purported hundreds of lions was at last in our hands. But most of all, I wanted that precious co-operation agreement, signed by environment minister Maria de Fátima Monteiro Jardim and her soon-to-be successor, Paula Francisco Coelo, and a letter of permission from the governor.

We were dead keen to get started: the research vehicle was packed and ready, and we had employed a highly skilled Namibian San Bushman tracker, Berry Alfred, to train counterparts in Angola. I traced the whereabouts of the brilliant field researcher and technician, Seamus Maclennan – holed up in his apartment in New York – and sent for him. The team was now ready and we hit the survey trails within days of the signing ceremony.

© Colin Bell

Since peace has returned to Angola, wildlife numbers have been on the increase. Species such as these southern sable (*Hippotragus niger niger*) are now regularly encountered in Luengue-Luiana National Park, and lion populations should rise over time, if poaching can be kept at bay.

Days turned into weeks and weeks into months. Seamus took his team and trusty Land Cruiser to parts of Luengue-Luiana National Park that few people had seen since before the civil war. Throughout the remaining dry months of 2015 they went where once landmines had been laid. I remember waiting for the team to return from each trip, hoping they were okay and waiting to hear about all the lions they had encountered. Fortunately, they always got back to camp or found a way to message me for more fuel and supplies. Sadly, reports of lions were few and far between.

Lions in camp

On one trip, I joined the team to deploy camera traps with a student from America, Carolyn Whitesell, who was finishing her PhD on human–lion conflict in the Okavango Delta, Botswana. She was no rookie but, along with her assistant Geraldo Mariya, protested when I suggested we sleep under the stars at a likely looking pan. Soon after I put my head on my pillow, two male lions roared nearby. Without a tent, I confess to moving my groundsheet up next to the car.

The next morning, we followed their tracks for many kilometres along a road we deemed suitable for camera trapping, placing traps every four kilometres, and then waited anxiously for the exciting results. The photos did not disappoint. We caught these truly magnificent Angolan lions a few times on camera.

Carolyn and Geraldo deployed another 200 or so camera traps, many of which were subsequently stolen or destroyed by meat hunters (poachers). But, having succumbed that first time, they never again slept under the stars without a tent. The next time I did so in Angola, two lions again visited the camp: it's my secret 'lion call-up' method.

Seamus, tracker Berry Alfred, Carolyn and Geraldo worked tirelessly through 2015 and for six of the dry months in 2016. During 2016, not a single sign of lions was found anywhere in either Mavinga National Park or Luengue-Luiana. The only lions hanging on precariously in the region were living close to the Namibian border. Seamus and his team never stopped looking, but the lions were just not there.

Meat hunters, no rules, no lions

Prey was heavily depleted and obviously being poached; meat hunting was not considered illegal. From poor peasants to army and police officials, everyone was hunting wild animals. And the lions had simply run out of prey. Even in areas where prey was abundant, adjacent to protected areas in Namibia, breeding-age lionesses were almost non-existent.

After two years of incredibly hard work and thousands of hours and kilometres searching for lions, we had too little data to compute a reliable lion population figure with confidence. The best we could do was estimate that about 30 lions still lived in a protected area four times larger than the Kruger National Park. It was disheartening.

Where were all the lions? Dead, is the answer; a long time ago, too. Hunted for sport by the Portuguese colonists and then starved or killed during the civil war, they were all but gone. Throughout these two parks, and in other parks in southern Angola that we surveyed (Iona, Mupa and Bicuar), we estimated that perhaps a handful of lions were left in Cangandala National Park and fewer than 30 in Luengue-Luiana.

Bleak, yes, but I relearnt what I already knew: lions and people do not live well together without strong conservation programmes and true and equitable community benefits. Also that wars kill lions: hungry soldiers and scared, even hungrier, peasants are not a good recipe for lion conservation. Africa can be extremely unstable politically at times: violence erupts and rural people and wildlife always suffer.

These scars and effects take a long time to heal. The shock of the demise of Angola's lions did, however, resonate with the international community. While I still ran the programme, we raised about US$2 million to start the first community-involved lion-conservation programme in Angola. I no longer run it and I understand that COVID-19 and other challenges have had negative effects on the programme.

Strong and good leadership, and the right people on the ground, are needed to save and recover the last of Angola's lions … in truth, everywhere in Africa. Rural people need to benefit from all forms of tourism and conservation activities. It is a real challenge and one, these days, I miss dearly.

25

Lions: quo vadis Angola?

The end of the civil war in Angola has rolled back a curtain on a most magical and precious wilderness. Effectively managed and protected, it could become one of the world's greatest havens for lions and much other wildlife besides.

Colin Bell

ANGOLA

For 30 years, central and southeastern Angola was the epicentre of a devastating civil war that pitted Angolans against fellow Angolans. Behind each ideological faction were the superpowers playing out their Cold War agendas with Russians, Cubans and East Germans on one side and South Africans on the other, with the Americans and British lurking in their shadows.

By the time the civil war ended in 2002, an estimated 800,000 Angolans had been killed, hundreds of thousands maimed, and four million had lost their homes and livelihoods. Many villagers were forced to seek refuge in the remotest parts of the country, with no alternative but to live off the land to survive. Soldiers who needed to be fed placed unsustainable pressures on wildlife populations. Armies had to be financed; ivory was traded for weapons, while land mines finished off much of the remaining wildlife. After the civil war, precious little wildlife survived. And with no prey, lions largely disappeared from most of Angola.

With the advent of peace, however, Angola started on the slow road to recovery and normality. The capital, Luanda, and the north prospered from their oil revenues. The giant sable antelope (*Hippotragus niger variani*), the national symbol of Angola, was thought to be extinct; but in 2004, camera traps in the centre of the country picked up that a few individuals had survived. Thanks to the work of Dr Pedro Vaz Pinto and his team, their numbers are again on the rise. Yet, little was happening in the southeast of the country. This part of Angola remained the great unknown.

Across the border in Botswana, safari guides at the time speculated around campfires about the source of the millions of cubic metres of pure clean waters that flooded into the Okavango Delta each year from 'somewhere' in central Angola. There were no weather satellites nor Google Earth, yet, to help inform the speculators.

Then peace arrived in 2002. The remarkable HALO Trust has been de-mining large tracts of Angola; over 100,000 mines were destroyed and the country began to open up. In 2015, a small group of hardy scientists and adventurers from the Wild Bird Trust set off on their quest to find the source of the Okavango Delta. Over a period of nine years, the Okavango Wilderness Project, in collaboration

One of the many 'source lakes' found in central Angola that are the headwaters of the Cubango/Okavango, Kwando, Zambezi and Congo rivers. These lakes and the peat beds that surround them feed pure, fresh waters all year to the landscapes downstream.

Midway between Menongue in central Angola and the Namibia–Angola border lies the Cuatir Nature Reserve. With flowing water year-round and fertile floodplains, Cuatir has the potential to become a world-class wildlife reserve. © Colin Bell

Angola is at the crossroads – will it take the high road or shuffle steadily down the slippery low road?

with *National Geographic*, has explored and documented a great and uncharted wilderness. Around 13,000km of *mokoro*-ing, walking, and even fat-biking later, the mysteries of these unexplored lands, forests, lakes, peat sponges, rivers and their tributaries are being revealed to the world. In the process, over 130 species have been discovered that are new to science and are being documented.

The team, led by Dr Steve Boyes, consisted of mainly Angolan and southern African scientists, highly skilled *mokoro* paddlers, and a few safari guides with a sense of adventure. The baYei *mokoro* paddlers from the Okavango, with their innate knowledge of the moods of the waterways, were key to the expedition's successes.

Thanks largely to the sweat and toil of these expeditioners, we know a lot more about the headwaters of the Okavango Delta. These extraordinary highlands are now recognised as a crucial water tower on Earth. The annual rainfall of about 1,500mm a year replenishes a labyrinth of lakes, vast peat bogs and sponges, allowing the water tower to relentlessly conduit fresh water into the arid lands, which are fed by the Okavango, Zambezi and Kwando river systems all year, even during the harshest droughts.

These rivers provide much of the water that sustains the Kavango–Zambezi Transfrontier Conservation Area (KAZA); an area greater than France with one of the planet's largest trans-boundary concentrations of wildlife anywhere. Close on 70% of all the savanna elephants can be found in KAZA and the area is regarded in the lion world as having the most significant lion clade.

On a more recent expedition, the exploration team followed the north-flowing rivers from the source lakes all the way to the Congo. Depending on where a drop of rain falls in these highlands, it could either end up in the Indian or the Atlantic oceans or dissipate into the Kalahari Desert 1,500km away.

Ironically, a 'benefit' of Angola's nearly 30-year civil war was that its insecurity and the dangers of war kept every possible user or polluter out of southern Angola. As a result, the entire Okavango system was fully protected. Around 26 million hectares of prime wilderness from the river's source in the highlands up north all the way to the Botswana border were free of any threats or development. During these civil war years, rivers flowed from the source lakes, year in and year out, delivering pure, fresh, uncontaminated natural waters to the Delta as they had done for millennia.

However, since peace has returned to Angola, change is in the air; everything is up for grabs. Chinese rosewood and teak loggers, commercial farmers, canal and dam builders, international frackers and miners are eyeing, plundering or profiting from Angola's natural resources. Most of these destructive impacts are happening in central Angola, to the west of the highlands and Okavango's source lakes, but they are steadily moving eastwards and it won't be long before developers start tampering with the precious flood waters that feed the Okavango Delta.

Already a small number of canals have been built, which tap waters from the Cubango River for large-scale agriculture projects. A development plan, published in 2018, proposed constructing many 100m-wide canals and tunnels, which would divert an enormous amount of water away from the Okavango for massive farming enterprises. And if any of that happens, what will happen to the Delta?

We can travel great distances in Angola today and not see signs of a single wild animal besides maybe an unlucky duiker displayed as bushmeat for sale along a roadside. However, as nature starts clawing its way back, rewilding is happening in a few places. Some of Botswana's elephants have followed their ancient migration instincts northwards and have started to repopulate southeastern Angola. In the Luengue-Luiana National Park and the Bico area in the southeastern corner, where Namibia, Zambia and Angola meet, wildlife numbers are climbing the fastest. Large herds of buffalo, common sable, roan and many smaller antelope species now complement the elephants. And now that there is ample prey, the lion populations have started to rise again. There is real potential for this area to develop into a prime wildlife destination, especially if the African Parks Network is allowed to bring in their expertise to manage the southeast sector of the country.

What is the future for the other, more remote, less inhabited wilderness areas around Angola? With up to 81 million hectares of potential lion habitat with low human densities, and with a planet desperate to mitigate against climate change, you would imagine the developed world would be throwing carbon money at Angola. Just the headwaters of the Okavango Delta in Angola cover close on 24 million hectares – more than five times the size of Switzerland. All this priceless, irreplaceable and sparsely inhabited land sequesters millions of tons of global carbon emissions in its soils and peats every year, and will do so in perpetuity if it is formally protected through a network of well-managed conservancies, parks and reserves funded by carbon offsets with full community participation and beneficiation.

Angola has the potential to become a global nature-based tourism powerhouse off the back of its vast wildernesses and their wildlife-carrying potential, but only if its tourism and wildlife sectors are carefully and sensitively developed. They are the late developers in this field, so enjoy the advantage of learning from the best practices and avoiding the pitfalls of the worst.

The hope and dream is for Angola to sensitively develop its wilderness assets to become a prime wildlife and tourism destination. It could once again be home to 500–1,000 free-roaming lions. That would be quite a turnaround! The Okavango Delta, too, would in turn be secure and the KAZA unifying dream, on which the presidents of Angola, Botswana, Zambia, Namibia and Zimbabwe signed off in 2011, would become a reality.

Angola is at the crossroads – will it take the high road or shuffle steadily down the slippery low road? The high road is a distinct possibility if there is the appetite within Angola and the developed world to change its antiquated carbon-offset policies, and be rewarded for setting aside large tracts of land for wildlife and wilderness forever. And if this rewilding happens, lions will be crucial beneficiaries on a continent where their sanctuaries are steadily becoming boxed in by rapidly increasing human populations.

A large herd of galloping roan antelope on the vast grassland savannas southwest of Cuito Cuanavale in Angola. With its vast areas of wilderness, Angola could become a haven for lions and other wildlife.

A relatively new 8,100ha agricultural farm – which can be viewed on Google Earth in the Menongue area (14° 44' 12S and 16° 32' 16E) – is extracting all its water from the Cubango River and thus contributing to the Okavango Delta's low flood regime these past years.

One of many new Chinese-owned sawmills. This one is located in central-southern Angola for processing the rosewood and teak forests that are being decimated in the region.

A mobile safari in Luengue-Luiana National Park, Angola.
© Stefan van Wyk

The Shire River, Liwonde National Park, Malawi. © Dana Allen

MALAWI

26

The lions of Liwonde

Under the stewardship of African Parks, lions are returning to this beautiful park in southern Malawi.

Chris Badger

The first wild lion I ever encountered was in Kenya's Maasai Mara in 1980. I was driving a five-tonne Bedford truck and we had employed a local guide. He sat on the roof and guided us through the long grass to a lone male lying asleep in the open, surrounded (even back then) by several minibus loads of tourists. We quickly realised our guide had not been looking for lions; he had been looking for minibuses. As it was early afternoon with the sun almost directly above us, the large truck cast only just enough shade for the lion to amble over and lie down right underneath us. It was a less-than-perfect introduction to such a magnificent animal.

I arrived in Malawi in 1987, having spent several years as a guide in Botswana and South Africa. The intensively managed and well-protected Sabi Sands in South Africa guaranteed regular sightings of lions and, in the glorious wilderness areas of northern Botswana, it required no particular skill or local knowledge to bump into them. Some of the sightings were extraordinary: territorial battles between huge males; large prides hunting buffalo and, occasionally, elephant; fights over kills with large packs of hyena; copious mating; close observation of the rearing of their young; and the wondrous memories of drifting off to sleep to the full-blooded roaring. It is impossible to come away from these experiences without an abiding fascination and respect for these most iconic animals. It was a privilege I simply took for granted. The thought that lions would soon become critically endangered never occurred to me.

On my first drive into Liwonde National Park in March 1987, I saw a pride of 19 on the magnificent Chikalongwe floodplain and, from 1988 to 1992, the lion viewing in the park was consistent. With regular visits on mobile safaris, and then getting the concession to develop and operate Mvuu Camp and Lodge, we had the opportunity to explore the area further and reckoned there were at least four resident prides and regular visiting lions from Mangochi Forest Reserve to the north and from Mozambique to the east.

Elsewhere in Malawi, there were small, isolated populations. Large nomads would wander in from eastern Zambia into the montane grasslands of Nyika, and Nkhotakota, Majete and Vwaza Marsh all supported small populations. There were even small prides living outside the protected areas near the southern shores of Lake Malawi and in the

The first generation of cubs born in Liwonde National Park after lions were reintroduced following a 20-year absence due to poaching and habitat loss. © Cheryl Jayaratne

Bwanje Valley. Living on the lake near Monkey Bay between 1987 and 1989, we regularly heard lions roaring from the hills to our south.

However, by around 1998, all the resident lions of Liwonde had been poached. This poaching was not well documented and few carcasses were found, but with a busy lodge and camp, we would have found them if they were there. We found no tracks and heard no roars. Occasionally, a lone male or a couple of females would wander into the park, presumably from Mangochi Forest Reserve or perhaps from Mozambique but this, too, stopped around 2010. The poaching was indiscriminate and probably not specifically targeting lions but the preferred mode of using wire snares on game trails inevitably caught them, as did some poisoning of the waterholes.

Malawi's challenges in preserving the integrity of its wilderness areas are not unique but they are perhaps greater than some of its neighbours, which have much larger protected areas and far fewer people. Malawi has a large and ever-increasing rural population. In 1987, it was around seven million. In 2020, it was around 17 million. The subsistence poachers of Liwonde are simply the sons and grandsons of communities that fished and hunted the area as a birthright and a means of survival. The rapidly increasing population has simply tipped the balance from sustainable to unsustainable. My view has always been that if Malawi can find a way to protect, nurture and sustain its remaining wilderness areas, it sends a powerful message to other countries with similar challenges.

A glance at a map will show how several of our wilderness areas were interconnected until relatively recently but that human population pressure has effectively narrowed or closed the corridors that connected them: Liwonde to the Mangochi Forest Reserve and further on into Mozambique; Nhkotakhota to Kasungu and on into Lukusuzi and the Luangwa in Zambia; and Nyika National Park and Vwaza Marsh into Zambia's Luambe and North Luangwa. Getting these corridors back is at best hugely challenging and at worst impossible. The net result is that parks such as Liwonde have to be intensively managed and protected as isolated islands of wilderness surrounded by rapidly expanding rural communities. Our wilderness areas are simply competing with other forms of land use. The connecting corridors cannot realistically be re-established with the notable exception of

Department of National Parks and Wildlife rangers in Liwonde, Malawi. © African Parks

Liwonde's link to the Mangochi Forest Reserve, which African Parks is now managing as a whole with the park.

Providing a habitat in which lions can thrive may seem elitist if the maize harvest fails and surrounding communities go hungry. However, Liwonde and other protected areas have a strong argument to remain protected. The Shire River on the park's western boundary provides most of the region's electricity through a large hydroelectric plant on the lower river, the riverine forest and mopane woodland is a substantial carbon sink,

Lions being reintroduced to Malawi.
© African Parks / Sean Viljoen

and there is potential for tourism to provide exponential benefits locally. In a survey we conducted in 2012, we found that one job at Mvuu supports from eight to 10 people through the generation of much-needed foreign exchange; and furthermore, the soils of the park are simply not suitable for agriculture. The other strong argument – which many protected wilderness areas can lay claim to – is that well-managed and protected areas can remain intact in perpetuity. The moral imperative is that we, as humans, have no right to treat wildlife and wilderness with the huge disrespect we often do. We must always temper our actions and consider these practical realities if we are to step back from the brink.

The arrival of African Parks

In 2017, African Parks took over the management of the park on behalf of the government of Malawi. The Department of National Parks and Wildlife had struggled for years with a shortage of resources and funding, and this was a far-sighted move on their behalf to guarantee the long-term future of the park. Very quickly the park's boundaries were protected by state-of-the-art solar-powered electric fencing, and a vigorous anti-poaching programme was put in place. This was followed by a series of reintroductions of cheetah, buffalo, black rhino and leopard. A series of lion reintroductions followed. In May 2018, two male lions from Malawi's Majete Wildlife Reserve (originally from Pilanesberg) and two females from Venetia in South Africa were introduced to Liwonde. In August 2018, two males from Dinokeng and three females from Shamwari and Lalibela arrived and, in March 2019, one female arrived from Majete (originally from Lalibela). All the animals were aged between two and nine years at the time of reintroduction.

Currently, there are 19 lions in the park: two adult males, six adult females, and 11 cubs. The Southern Pride consists of an adult male, four adult females, and seven cubs (three born in March 2021, and four in June 2021). The Northern Pride consists of one adult male, two adult females, two cubs born in September 2021 and two more born in October 2022.

Early signs are extremely promising: territories are being established, breeding appears to be successful and the overall aim of establishing the correct demographic population of lions within the park's 600km² is on track. It will require careful management and monitoring to achieve a long-term stable population but once more, visitors to the park can drift off to sleep to the roar of the lions.

Shire River floodplains, Liwonde National Park, Malawi. © Danny Badger

MALAWI

27

Rewilding Majete

Majete Wildlife Reserve encompasses 70,000ha of towering granite hills, grassy plains and miombo woodlands set in the Lower Shire River valley in the southwestern corner of Malawi.

Africa Geographic

The Shire River is the only outlet of Lake Malawi and winds its way through central and southern Malawi and the Majete Reserve before eventually joining the Zambezi River to the south. Bulbous star chestnuts have forced their way up through the reserve's rocky kopjes, and hulking baobabs dot the savanna, while the riparian forests of the river valley are lush with ilala palms and ancient mahoganies.

Every river in Africa has its own distinctive character, and these landscape lifelines tend to dominate and define the continent's wild spaces. So it is in Majete, where its two main rivers are entirely unalike. The Mkulumadzi is gentle and charming, flowing through patches of boulders and along small open floodplains. The more famous Shire is mighty, fast, forcing its resident hippos and crocodiles to endure the currents.

The Shire subsumes the smaller Mkulumadzi in the heart of Majete and continues south to tumble over the picturesque Kapichira Falls. Nearly 200 years ago, David Livingstone and his crew followed the course of the Shire from the Zambezi River upstream and through Majete, which would eventually lead them to the shores of Lake Malawi.

Second chances and new hopes

Majete Wildlife Reserve was the first protected area to fall under the management auspices of African Parks back in 2003. Prior to that, the reserve was a reserve in name only. Charcoal burning and logging had divested vast patches of land of their woodlands, and poaching had claimed the lives of everything from elephants and rhinos to the once vast herds of buffalo and innumerable antelope species. Not one tourist had ventured to the reserve in three years. After a lengthy negotiation, the non-profit organisation African Parks was invited to enter into a public–private partnership with the Malawi Department of National Parks and Wildlife

234 ~ THE LAST LIONS

The Shire River, Majete National Park, Malawi.
© Marcus Westberg

(DNPW). As part of their now well-established (but then trailblazing) method, African Parks assumed responsibility for the rehabilitation and long-term management of the reserve, as well as for all the costs, while providing a source of sustainable income for surrounding communities. They inherited a 'wasteland with no perceived value, and little to no hope for a revival'.

But revive it they did, in a careful and pain–staking process that has taken close to 20 years. Rather than rushing into a hasty attempt to restock Majete, management teams set about establishing the necessary infrastructure and relationships needed to secure the reserve's long-term future. As is fundamental to the African Parks' approach, the first step was to institute community development and engagement programmes to prove the importance of protecting the reserve. Fences sprang up around the perimeter, the road network was increased tenfold, and lodges, camps and a visitor centre were prepared for the inevitable return of the tourists.

The establishment early on of a smaller and more manageable fenced inner sanctuary (since dismantled) provided a safe haven for both the remaining and translocated wildlife to flourish. Meanwhile, the reserve's team of 12 grew to over 140 well-equipped, well-motivated and well-trained staff members. In a remarkable testament to their efforts, not one elephant or rhino has been lost to poaching since their return to the reserve.

Tourists have begun to reappear in their droves, with 13,000 visitors generating more than US$740,000 in revenue in 2023. Aside from the reserve running costs, this money is channelled into community education scholarships, social infrastructure such as clinics, malaria prevention, and beekeeping projects.

Back from the brink

With remaining wildlife numbers on the rise and their safety ensured, the time had come to bring back the species lost to poaching. First came seven black rhinos in 2003, followed by the arrival of a herd of 70 elephants in 2006 and additional herds in 2008 and 2009. Today a healthy population of elephants roams Majete Wildlife Reserve, fulfilling their role as ecosystem engineers and reversing the damage to the reserve's habitats. Translocated giraffes, buffaloes and sable antelopes all joined the ranks of a burgeoning herbivore population. Since African Parks took over management, more than 5,000 animals from 17 species have been brought into Majete.

Today the reserve is home to a multitude of antelope species, including waterbuck, eland, Lichtenstein's hartebeest, common duiker, kudu, nyala, bushbuck, Sharpe's grysbok and suni. With sufficient prey species on the menu, the large carnivores were next on the arrival list. Lions, leopards, cheetahs and, most recently, wild dogs have all made the journey from other countries (mainly South Africa – a trip of over 5,000km) to restore Malawi's predator populations. Today, Majete is home to a thriving lion population with 62 known individuals and an estimated total of around 80. These animals fall under the umbrella of the Malawi Predator Metapopulation Management Plan, allowing for an overarching management approach. Along with elephants and rhinos, they and their offspring have aided in the rewilding and genetic supplementation of other reserves around Malawi, such as the Nkhotakota Wildlife Reserve in the north of the country. The gradual rehabilitation of Majete has also been of tremendous benefit to its avian residents (and migrants). Over 300 bird species have been recorded in the reserve.

There could be no better description of the restoration of Majete Wildlife Reserve than a labour of love, born of 20 years' worth of dedicated toil. In many ways, for African Parks, it also set the stage for future projects, serving as a platform to trial new and innovative methods to protect the continent's most vulnerable wild spaces and prove just what can be accomplished. What's more, Majete's success became a springboard for Malawian conservation, echoes of which can be seen across the country's other parks and reserves. As a result of the success of Majete, African Parks entered into management agreements for Liwonde National Park and Nkhotakota Wildlife Reserve, both in Malawi, in 2015.

African Parks offers safari camps (lodges and campsites) where 100% of tourism revenue goes towards conservation and addressing the needs of local communities.

Majete National Park, Malawi. © Marcus Westberg

Busanga Plains, northern Kafue National Park, Zambia. © Michael Poliza

The lions of Zambia

PORTFOLIO

Busanga Plains, northern Kafue National Park, Zambia. © Dana Allen

Busanga Plains, northern Kafue National Park, Zambia.
© Dana Allen

Lions on the prowl near Mfuwe Lodge in South Luangwa National Park, Zambia. © Isak Pretorius

ZAMBIA

28

Learning with lions

After being caught in the gaze of an angry growling lion only metres away, I – a young Zambian – was inspired to dedicate my life to saving them.

Thandiwe Mweetwa

It was no ordinary day – 28 May 2009. On that day I had a staring contest with a lion. For what felt like an eternity, I remained crouched in the grass, a few metres away from a 200kg growling, young male lion. It was my first day of work as a field research volunteer for an organisation called the Zambian Carnivore Programme. That day, while accompanying a team of researchers on a mission to collar a male lion, I experienced something that cemented my decision to join the wildlife sector.

I got the opportunity to experience the full power of three young male lions roaring within five metres of our vehicle. I had never experienced anything like it before. I knew I was in the right place and that I wanted my career to involve conserving lions.

My story starts in Mazabuka, a small sugarcane plantation town in southern Zambia where I spent my childhood. My mother was from Mfuwe, a small village just outside South Luangwa National Park, one of Zambia's premier wildlife sanctuaries. Her voice would fill with pride, nostalgia and a bit of homesickness when she told us amazing stories about the incredible wildlife that surrounded her village far away. My fascination with nature was nurtured by those elaborate, sometimes exaggerated accounts. Repeats of nature documentaries watched through our 12-inch black and white TV brought my mother's stories to life.

After my parents passed away, I moved to Mfuwe with my family. It was a difficult time of transition but I had the opportunity to live close to wildlife and learn more about it. It encouraged me to pursue a career in conservation. I wanted to become a wildlife veterinarian so I could treat animals that had been injured by wire snares. My dream was on course until I got to the anatomy and physiology classes in my second year of university. A few lab sessions in, I discovered working hands-on with animals was not for me. I came back home that summer, feeling dejected. I had worked so hard to get to this point.

Fortunately, a new organisation, the Zambian Carnivore Programme (ZCP), had just started operating in our area and its primary focus was studying the population dynamics and threats facing carnivores in Zambia. They were looking for volunteers, which is how I found myself in the middle of Mtanda Plain having a staring contest with a lion. It was my first up-close encounter with the species and I realised I did not have to be a vet

246 ~ THE LAST LIONS

Luangwa Valley, Zambia. © Dana Allen

to work with lions and other carnivores. Instead of being killed by that young lion (it simply broke eye contact and walked away), it was the beginning of what has been a challenging yet immensely fulfilling career in wildlife and large carnivore conservation.

Lessons from lions

Since that magical day, I have had the privilege of knowing and working with many lions. Some of them were tiny cubs when I first met them and are now old lionesses well into their twilight years. From observing these beautiful animals, I have learnt courage, perseverance and resilience. Many standout episodes have demonstrated the characteristics that make their kind such global icons.

An example of this was a favourite lioness of mine, nicknamed Chimozi by local safari guides. Her name meant 'only child' because she was the sole survivor out of a cohort of seven cubs. The rest were killed in a series of territorial disputes between two sister prides in an area called Lion Plain. Following the final battle between the two groups, Chimozi, still a cub, was temporarily abandoned by her pride in their bid to escape after being outnumbered and overpowered.

She wandered the scorching plains alone until she was reunited with her group several days later. The bush is a dangerous place for a lion cub, even within the fold of a protective pride. Surviving alone in a hostile territory patrolled by a super pride that had killed six of her cohort mates was a remarkable feat. This successful solo trek made Chimozi a poster child for resilience.

Over on the eastern side of the Luangwa River in Nsefu Sector lived another one of my special lions. She was a slight, ageing lioness nicknamed Ambuya ('grandma'). Her instincts as a mother were lacklustre but her tenacity and skills as a hunter were unparalleled and, even in the later years of her life, she remained a formidable predator. As soon as night fell, Ambuya transformed from a frail-looking lioness into the leader of her hunting party, taking dangerous prey head-on. Throughout her long life, she survived many hunting injuries, including a buffalo kick to the face, which left her with a misaligned lower jaw.

Another lion that stood out was a beautiful blond male named Cutthroat. I first saw him as a small cub stashed in a Combretum bush. Over the years, he morphed into a strong young male ready to set out and establish his own pride. Unfortunately, a few months after he left, he was caught in a wire snare that nearly ended his life. We were able to rescue him and help him on his way to recovery. Cutthroat and his coalition brothers eventually set up their own territory and went on to have cubs. His journey highlighted, in painful detail, the challenges young males face when they leave a pride to establish their territories.

Lions in a changing landscape

The Luangwa Valley is home to Zambia's largest population of lions and wild dogs. Although it is one of Africa's last remaining strongholds for lions, the area has not escaped many of the challenges that face Africa's ecosystems. The valley is affected by habitat loss and the impact of the illegal bushmeat trade, which has resulted in reduced prey for lions and other carnivores. To understand the effect of human activities on the species, I focused in my master's degree on the response of the lion population to a temporary ban on trophy hunting.

As with many parts of Africa, our beautiful landscape has undergone many transformations. Historically, the Luangwa Valley was inhabited mostly by hunter and agriculturalist tribes. Livestock rearing was not common. With the high incidence of tsetse fly, which transmits trypanosomiasis to livestock, many domestic animals that came into the valley fell victim to the disease. Thanks to recent changes in the vegetation and other tsetse fly control measures, it is now possible to keep livestock, so people from the plateau have been moving into the valley in search of greener pastures. There have also been several alternative livelihood programmes supporting livestock farming as a way of generating income and boosting food security in the flood-prone area. This has increased cattle numbers and amplified conflict between humans and carnivores, especially lions. Human–wildlife conflict is common in many parts of Zambia and other areas of Africa, but is new to our area, forcing us to address the issue. Reducing this conflict is critical for lion survival as problem lions tend to be shot by the authorities or killed in retaliation by communities.

There have been incidences where entire prides have been wiped out this way. In one such incident, the world-famous Marsh Pride was poisoned a few years ago and, more recently, around eight lions were speared to death at a conservancy in Kenya by irate community members.

In a career that has spanned nearly 15 years, I have seen how sharing the land with lions brings as many costs as pleasures. A core of my current job involves working with communities on how to address human–lion conflict and the problems range from livestock predation to injury and death.

One particular evening showed me how uneasy the relationship can be between people and wildlife. Between 2019 and 2021 there was a man-eating lion ranging in the game-management area just outside South Luangwa Park. It was suspected to have attacked 11 people, five of whom were killed.

For conservationists, it was a difficult time. In September 2019 we had to face the family members of someone who had just been killed. When we arrived at the scene, we found an irate mob of some 30 people. They were yelling at us for taking too long to respond, saying we cared more about the lion than their dead relative. The victim's partially consumed body was still on the ground, though I didn't have the nerve to inspect it. More experienced team members worked to placate the crowd and it took several hours.

The crowd demanded that the lion be hunted and killed before allowing the victim's body to be moved the morgue. Permission was granted to do this because it is believed that once a lion kills and eats a human, it is likely to do it again. However, though we had a brief sighting of the culprit, we were unable to hunt it down and kill it. We eventually managed to calm the people enough for them to allow us to transport the body to the morgue. It was a lesson in how difficult it is to build harmony between people and lions.

In the months that followed, we did everything we could, working with professional hunters and various community members, to find the man-eater, a clever male in his prime that managed to escape all the baits and traps that were set for him. He went on to kill three more people and injured two others before eventually going quiet and disappearing about two-and-a-half years after his initial attack.

Striking a conservation balance

It's clear that we have to balance conservation goals for charismatic species such as lions with the development needs and wellbeing of the communities that live with them. A significant number of lions live among people outside of protected areas, with consequences for all. As we implement conservation efforts to repopulate lions in areas where they previously existed, we need to work out – and invest in – ways to make co-existence feasible and to reduce the cost of living with wildlife.

Unfortunately, though it's widely agreed that promoting better coexistence is a conservation priority, there is a massive shortage of funding for human–wildlife conflict (HWC) projects. One just has to look at the vehicles and resources allocated to anti-poaching work in comparison to other community-based projects: it is easy for communities to assume animals are considered by authorities as being more important than people. Organisations have to be very creative in order to raise the finances needed for HWC mitigation work. If we are to secure a future for lions, this has to change.

Since 2019, the ZCP has been developing a conflict-mitigation programme, working with various departments and agencies. With partner organisations, we formed an HWC working group that meets regularly to discuss the challenges and how best to address them. At ZCP, we have an HWC team comprising a co-ordinator and three assistants. Their task is to educate community people about personal safety and safe herding practices as well as encouraging them to learn more about lion behaviour. To limit the predation of livestock by lions, our team has built demo predator-proof enclosures, especially within the conflict hotspots. Through this work, it is clear that collaboration is the way to go.

Zambia is a huge country with nearly 30% of its land dedicated to conservation. There is great potential for lions to thrive here. Despite current challenges, I'm hopeful people are still willing to coexist with these predators. For now, lions encountered in community areas are generally still tolerated, even when they repeatedly cause conflict through livestock predation.

Liuwa Plain National Park, Zambia. © Andrew MacDonald

ZAMBIA

29

Life with Lady Liuwa

She was a lion alone after her pride had been shot and the only creatures who didn't run from her were us, so we became her family. It was both an honour and a discomfort.

Andrea & Craig Reid

We first met Lady Liuwa in 2007 when our family moved to Liuwa Plain National Park in western Zambia. My husband Craig was the new park manager and I spent time trying to teach our children Kyle and Erin, aged six and three, as well as assist Craig in the park. He had been to Kalabo and Liuwa before and was told about this legendary lioness that had once been part of a pride but now spent her days roaming the vast plains alone.

Staying in Kalabo in those days was hard and initially a huge culture shock. We had lived in Hluhluwe-iMfolozi Park in South Africa and were accustomed to a remote life, but Kalabo made it seem like we had been living in a modern city. After a six-hour drive from Lusaka to Mongu, we would hop onto a speedboat and, three hours later, arrive in Kalabo. During the dry season, we could take the road from Lusaka across the Zambezi floodplain to a pontoon, which ferried us across the Zambezi River, then back onto another bumpy road to Kalabo.

I don't know which route I loathed more, the trip across the floodplain to the pontoon, which was sometimes broken or had run out of fuel – or the three-hour boat trip; during the wet season, we invariably got drenched. For parts of the year when the park was flooded and inaccessible, we stayed in Kalabo, a small noisy Lozi village. A missionary family there told us that Lady Liuwa had once been part of a pride that had been killed.

In 2003, when African Parks assumed management of Liuwa in partnership with the Department of National Parks and Wildlife and the Barotse Royal Establishment, wildebeest and zebra were in steep decline and grasslands were threatened by rice fields. Lady Liuwa was the last lion. We were told her pride had been shot illegally by a hunting outfitter operating in the game management area. Many years later, when we were working in the north of Zambia for a different project managed by African Parks, a hunter boasted that it was he who had hunted out the last lions of Liuwa.

The first time I encountered Lady, I was filled with anxiety. She had grown accustomed to humans and was completely unafraid, frequently coming to within a few metres of people in the camp. She spent most of her time in the southern parts of Liuwa, around the woodlands of Matiamanene

Liuwa Plain National Park in Zambia was floundering before African Parks became involved in 2003. Migration numbers in the park had plummeted from an all-time high of 60,000 wildebeest to around 10,000. Matters on the ground were gloomy for both people and wildlife. African Parks introduced their recipe of ample capital, world-class management and a policy of working closely with neighbouring communities, and today Liuwa boasts Africa's second-largest migration of wildebeest, with a population of over 40,000 and climbing. © Andrew MacDonald

Camp. She liked to lie near the campfire, next to your tent or even on its veranda. She was a beautiful lioness with a distinctive dark spot on the left of her muzzle, below her nose.

My encounters with her got better over time. I was able to rationalise that she had no evil intent and that we, the humans responsible for her aloneness, should not run away from her like everything else did and were her only company. Staying at Matiamanene, she would announce her arrival each evening by running into camp with low grunts, greeting us and perhaps explaining how her day had gone. Needless to say, we never let our guard down. Instinct is a survival skill and if we ran or approached her, we would become the hunted.

Working trips to Matiamanene, where Craig and I regularly hosted donors to the wildlife cause, were sometimes tricky. We had to take our children with us, which meant enforcing some rules. Our children were somewhat feral. They owned shoes but seldom wore them, they had nice clothes but these were reserved for rare trips to 'town'. The sand at Matiamanene was fine and had black organic matter in it that would turn feet black. For reasons I cannot explain, our children loved to play in it. When not doing that, they would rush about and climb trees. However, there was one rule that was never broken and strictly enforced. Come dusk, both children had to be inside the management shed.

During these years, I became a master of classic DVDs, bribery and corruption. We needed the children to be glued to the screen so they did not follow me outside while I was shuffling to and from the main dining area. On one occasion, Lady and I bumped into each other on 'our' way to the dining area: me to deliver snacks and drinks and Lady to take up her spot on the outskirts of the firelight. She was often so close we could hear her breathing. This may sound romantic from afar but – close up – it is not! She was a lioness and needed other lions to answer her call, to roll with her in the black dirt, share a meal, scrap and growl at – not humans around a campfire.

Lady spent most of the year in the south but as we found out later, during the rainy season she would move north and disappear for months. We hoped she would return with another lion but never did. No lions were ever reported there, not from scouts on patrols nor from any of the many monitoring flights that Craig conducted. She was literally the Last Lion of Liuwa.

Before we joined African Parks in Liuwa, we had worked with lions, darting and monitoring them. We knew their behaviour and ecology fairly well and it was sad to think she would spend her life without another lion. We would often talk about the possibility of introducing more lions to Liuwa and, eventually, it became a reality.

We received the go-ahead internally from African Parks head office and Craig set about getting the necessary approval from both the Department of National Parks and Wildlife and Barotse Royal Establishment. After what felt like an age, we obtained approval to catch one male lion from Kafue National Park and to dart Lady and fit her with a tracking collar. Craig enlisted the help of wildlife veterinarian, Dr Ian Parsons, for the darting and to fit the collar. It went routinely and was the first time we were able to physically touch her, which we did with great respect and fondness. We were concerned this procedure would negatively affect her relationship with humans and were relieved to hear her coming into camp a few nights later, apparently unfazed.

Craig and Ian then drove to Kafue National Park where they were told there was a suitable male lion that had no pride. Three days of searching later, they found him lying next to the Kafue River. He was darted with a sedative and loaded into the back of a closed-canopy vehicle.

The plan was to check on him every hour and to roll him over to minimise any fluid build-up from lying in one position for too long. At a fuel station, Craig jumped into the back and grabbed the lion's skin by the flanks and stomach to move him. The lion, blindfolded, lifted his head, let out a huge grunt and swatted Craig across the face with his paw, grazing him. The onlookers all ran away while the vet quickly administered the top-up sedative before exiting the vehicle with haste.

After 11 long, hot hours and many fuel stops, the team arrived in Liuwa. The lion was quickly offloaded into a boma in a thicket that Lady frequented, and we waited for him to come out of his sedation. As he woke up and stumbled towards a carcass to feed, it was clear that, despite the tranquiliser given to him, the new environment was stressful. Later that night we heard lots of growls

"Bon Jovi," the renowned dominant male lion and 'king' of Liuwa for many years, was relocated to Kafue National Park after venturing beyond the park's boundaries and killing livestock far outside his usual territory. Fortunately, his radio collar allows for continuous monitoring of his movements. His move to Kafue, located 300km east of Liuwa, is a positive development for lion genetics, as Bon Jovi has fathered so many cubs that a significant portion of Liuwa's lion population carries his genes.

© Andrew MacDonald

and what sounded like him hitting the electric fence and getting shocked. We left at first light the following morning and found the fence badly damaged. We switched off the electrics but our presence repairing the fence distressed him. We had no choice but to sedate him again and, sadly, he died in the process. We were all devastated.

Despite the tragic first attempt, we knew we could not give up and requested permission to capture a coalition of young male lions. We strengthened the original boma electrics and enabled areas where we could observe them without their seeing us. We agreed that the most suitable option would be to transport the lions in a reinforced antelope truck, awake rather than sedated.

The new date chosen was the middle of the wet season. We reasoned that, with flooded rivers, plains and abundant wildlife, any lion would not want to leave such easy pickings. A bonus was that they would have a water barrier preventing them from finding their way back to Kafue. However, going by road from Mongu was not possible. Fortunately, the Ministry of Health had recently offered a speedboat, which was fast with lots of space, and exactly what we needed.

Craig and the team returned to Kafue once more, but this time, since it was the wet season, they found it difficult to find suitable male lions. They spent many days hunting, and eventually found what they were looking for. Two males were darted and loaded into the truck where they were able to wake up from the sedative. The team drove through the night, getting to Mongu harbour at first light where the speedboat was waiting. The trip from Kalabo to the boma took roughly an hour. Everyone retreated to the secure area and anxiously waited while the males woke up. That night, we could hear cries from the boma as they tested the electric fence. I could see the stress written all over Craig's face and the next morning he left before dawn for the boma. He was delighted to discover they were okay.

We were not the only ones who heard their cries. Lady had come to investigate. Instinct told the males they should be somewhat aggressive towards this older female and Lady reciprocated by charging at the fence and then scent marking around the enclosure. But she stayed at the boma day and night, moving away only to hunt. Eventually, both lions began to accept her and she them, lying close to one another on either side of the fence. On the fifth day, the lions in the boma were nowhere in sight – they had escaped. But their collars told us they were still in the area and they were spotted together with Lady Liuwa. Liuwa Plains had a lion pride at last.

Many questions remained: would Lady mate with one of these males, would she raise cubs of her own and teach them how to live among the Lozi people living in Liuwa? The one thing was for sure, however: Lady was once more a complete wild lion with a non-human pride.

Liuwa Plain National Park, Zambia. © Andrew MacDonald

Liuwa Plain National Park, Zambia. © Inki Mandt

Liuwa Plain National Park, Zambia. © Ariadne van Zandbergen

Women from nearby local communities use portable fish traps in the shallow wetlands of Liuwa Plain National Park, Zambia, where they are permitted to fish during specific times of the year using traditional methods. © Andrew MacDonald

Community involvement
Community fishers in the park

Part of African Parks' successful conservation recipe is to make sure that the surrounding local communities are very much included in their conservation work in all the parks and projects they operate throughout Africa. Each project and community has its own needs and nuances. Here in Liuwa Plain, the local community are allowed to fish in the park in their traditional way at certain times of the year. The Barotse Royal Establishment is represented by two members on the African Parks Zambia Board to provide a voice for the people regarding governance and decision making.

Today, Liuwa Plain is the largest employer in the region, providing critical educational and health benefits to the communities. With the implementation of fishing permits, local people can sustainably fish within the park at many of the hundreds of traditional fishing pools. In 2022, new classroom blocks and teachers' housing were built, and 25 community teachers' salaries paid. The park's school sponsorship programme provides bursaries and covers schooling and boarding fees, along with textbooks, extra learning material and uniforms for over 200 children a year. There are now over 200 Farmer Field Schools with over 4,700 farmers and 130 beekeepers learning sustainable farming methods. These agricultural projects are having a positive impact on food security and reduce the need for local people to survive by poaching, while the overall engagement and direct involvement of the community in the Liuwa Plain project has helped to renew their commitment to the custodianship of this landscape.

Niassa is open to adventurous travellers looking to explore an unspoilt corner of Africa, staying either at Mpopo Trails Camp or even camping wild.

MOZAMBIQUE

30

Understanding coexistence

Coexisting with lions is dynamic, fragile and built into the human condition through a shared history. Destroy that connection and an ancient fabric of understanding falls apart.

Dr Colleen Begg, Keith Begg & Dr Agostinho Jorge

It's September 2022 in a rural village of around 2,000 people inside northern Mozambique's Niassa Special Reserve. Mama Bibi Amisse holds everyone's attention under the mango tree. She is a natural storyteller. With only a few prompts from Eusebio, the head of our lion-monitoring team, she tells of the night, back in the 1990s, that her husband was attacked by a lion and she rushed to his defence.

He was sleeping in a *chilindo* (an open thatch structure with a roof but no walls) in the fields. The lion stalked him and dragged him out into the bush. Realising her husband was being attacked, she took a piece of firewood and chased the lion off. By rallying the help of neighbours, she was able to carry her husband to the village clinic to receive emergency care.

But the lion was not giving up so easily and followed them. A group of men mobilised to go out and hunt it, initially without success. It was time to call on the help of the traditional healer who could intervene with the ancestral spirits. Finally, the lion was killed. This is coexistence, multi-layered and complex, with no hero and no villain.

Mama Amisse speaks of the terror of the night and the extraordinary fact that they are both alive today. The details are vivid, the audience rapt, and 30 years have not dulled any of the horrors of an attack at night by a predator. The stories of lion attacks are told and retold, part of the fabric of the past and the future. They become woven into the tapestry of truth, fables and sorcery that characterise any community deeply connected to a place and its wildlife. Living with dangerous predators like lions is part of our earliest human history – eastern and southern Africa were the evolutionary cradle of the modern lion, where they coevolved alongside us.[1]

This particular meeting was part of a workshop on lion conservation. Our aim was to share ideas and insight by bringing together those from the local area with long-lived lion experience and those with learned experience of lions from university training. There is no doubt lions are in trouble here and are declining across northern Mozambique as they are elsewhere in Africa. In Niassa, this decline is largely the result of snares set for wild meat that inadvertently catch lions and kill their prey. In the past few years, there's been a worrying escalation in the international illegal trade in lion parts, teeth and claws. Their future here lies in the hands of the people who directly share space with them.

Niassa Special Reserve, Mozambique. © Niassa Carnivore Project

Sprawling wilderness

Niassa Special Reserve (NSR) in northern Mozambique, on the border of Tanzania, has an unusual history. This sprawling wilderness of 42,200km^2 supports globally important populations of nearly 1,000 lions, at least 350 African wild dogs and a good population of leopards and spotted hyenas, as well as around 60,000 people in more than 40 remote rural villages inside the reserve. It is one of the only large, protected areas in Africa where people were never forcibly removed when the reserve was first proclaimed by the Portuguese in 1954.

There have always been people living here. This is in sharp contrast to most protected areas in Africa where fortress conservation with its deep colonial underpinnings is the norm.[2] Most people are farmers and fishermen practising subsistence, shifting agriculture, with crops planted in the wet season from December to April. During the long dry season, they have to live off what they harvest, with some fishing and honey gathering to fill the gaps. Many used to hunt with spears and traps, but this became illegal when the reserve was proclaimed.

Most protected areas routinely separate society from nature (and lions) with well-defined boundaries and rules, and only a special few are allowed to enter as tourists or researchers. What better place than NSR, then, to examine coexistence; what it is and what it isn't in a place where people have always lived.

When we arrived in 2003 and set up the Niassa Carnivore Project in partnership with the National Administration for the Conservation Areas (ANAC) and the reserve management authority, we were still entrenched in our southern African upbringing of fortress conservation. We (Colleen and Keith) had a smattering of experience in community conservation from Zimbabwe, and extensive experience working on conflict with carnivores in Zimbabwe and South Africa.

The next 20 years in northern Mozambique were a time of learning, stretching and rethinking conservation. Agostinho joined the team in 2008 and together we started to unpick what was learnt from what was real in this complex landscape. How could the large carnivores (lion, leopard, spotted hyena), which are so dangerous to livestock, and the growing population of people who live here continue to have a shared future? How do we give people a choice? What were the main threats to lions and people and what were the opportunities? Whose voices were missing?

Learning curves

Seeds of thought germinate slowly. One day in 2005 we were following a radio-collared lion designated LICM01 (Lion Capture Male 1), and locally known as Campo. He was the first male lion we had collared in Niassa so that we could follow his and other lion movement across this landscape. It was hard work keeping up with him by vehicle through Niassa's miombo woodland, mountains and rivers. At last, as the mid-morning heat set in, he flopped down under the shade of a huge boulder and went to sleep.

This is a distinctive spot and a landmark that's visible for miles around. While granite inselbergs or 'island mountains' are a signature feature of Niassa, rising unexpectedly out of the woodland, boulders like this one are not. It was too far to go back to camp, so we waited with him until he was ready to start moving again. Years of following animals have made us good at waiting; we rest, read and keep a lookout.

We noticed some red marks on the rock above the sleeping lion. Through binoculars, they appeared too regular to simply be old wasp nests. For years we had been looking for rock art on the granite inselbergs, but with no luck. Could these be paintings? At sunset, Campo started yawning, roared and eventually ambled off. We let him go and scrambled up to the rock face. The marks did look like paintings but were different from the San rock art we were more familiar with. These were way more abstract: fingerprints, dots, lines and circles.

The Bradshaw Foundation and Rock Art Research Institute would later confirm these were likely the work of Batwa hunter-gatherers, with northern Mozambique their southern limit. Their age is unclear, but the paintings are consistent with the red geometric art tradition that seems to have historically extended back thousands of years.

The image of a lion lying under this rock art is a graphic illustration of the notion that wilderness includes people.

The image we took that day of a lion lying under the geometric rock art is a once-in-a-lifetime moment and the perfect illustration of an essential point: wilderness includes people. This is a place still crisscrossed with pedestrian paths and sacred sites, home to generations of predominately Yao and Makua people since the 1600s.[3] It's unlikely that any part of this landscape has been untouched by humans making a living from the land – honey gathering, hunting and shifting agriculture – as well as the long history in the region of trade in ivory, slaves, gold, lion parts and more.

There is really no such thing here as a pristine wilderness without any trace of human influence. These paintings reflect Niassa's people and their deep connection to the land, the environment and its lions, as well as serving as repositories of indigenous knowledge critical for long-term conservation.

Night attacks

In early 2005, we heard of lions that had attacked and eaten people in villages in the Negomano district in the northeast of Niassa Reserve. At the same time, research from Tanzania noted a spate of attacks there. Another myth was overturned – it's not only old or injured lions that attack people: lions are opportunistic and, once they have attacked a person, they may well return to repeat the act. And sometimes prides develop a culture of man-eating.

Between 1990 and 2007, in southern Tanzania, lions attacked over 1,000 people.[4] Studies showed the attacks tended to be highest in the districts with the highest abundance of bushpigs and lowest abundance of natural prey. Bush pigs increased the risk of a lion attack by being nocturnal crop pests that force people to sleep in their fields in makeshift temporary shelters in the wet season and chase the pigs away.

During the night, these pigs are often an important prey species for lions, especially in areas where other more common lion prey are dwindling. For this reason, most attacks occur when people are tending their crops close to their village.

We decided to survey every village inside the reserve to better understand the conflict with lions, and we are continuing to collate the data 18 years later. Does it follow the same patterns? With hindsight, our initial questions lacked depth, but that was a lesson we learnt only later. Answers always depend on the questions: we are biologists, not social scientists.

We asked: Do you know of anyone who has been killed or attacked by a lion? What was the person doing before the attack? Was the lion killed? Where did the attack happen?

Back then we didn't ask the more important questions: How do you see lions? Why do you think they attack humans? What does an attack mean? How do you feel about lions?

As this data came in, showing 103 recorded lion attacks between 1980 and 2022 inside the protected area, we began to see patterns similar to those in southern Tanzania. Attacks are mostly at night when people are at their most vulnerable and lions are more confident. Most (more than 80%) happen either in the croplands or the village and very seldom in the bush while people are fishing, harvesting or walking to sacred sites. Nearly half the victims were asleep in temporary shelters in the fields or out on their veranda at night.

It is clear that suitable lion prey, such as bushpigs, are drawn to the fields at night; and it is far easier for lions to hunt here than chase after wildlife that's widely dispersed in the dense wet-season vegetation. With a diverse concentration of prey animals and relatively good visibility, agricultural croplands offer clear advantages to lions. Walking in the fields at night to chase crop pests and elephants makes people vulnerable. Although it is mainly adult men

Many local communities in the Niassa region now coexist with lions. In order to avoid attacks while sleeping, some locals build their bedrooms high off the ground, which helps reduce the risk of nocturnal encounters.

that are attacked, women and children are just as vulnerable; anyone who walks alone outside at night is potentially a victim in the eyes of a lion.

Mama Amissi confirms that she and her husband were habitually in the fields at night in the wet season to protect their crops from bushpigs, as well as during the day, from baboons, warthogs, elephants and kudu. They would sleep in the open. Surprisingly, she does not blame the lion or even appear to hold a grudge. She never suggests that life would be better without lions and the idea of getting rid of all the lions doesn't feature.

They should have been sleeping in a *sanja* or safe house on stilts, she says. But she has changed her behaviour. Since the attacks, she always takes an empty plastic container into her *sanja* at night so she does not have to walk in the dark to go to the latrine.

Traditional knowledge

The elders have always known how to live with lions and been aware of the dangers of sleeping on the ground. When the Jesuit priest António Gomes was shipwrecked off the east coast of Mozambique around 1645, he and his Portuguese crew went into neighbouring Makua villages to search for supplies. He noted huts in a village that were raised off the ground and the villagers explained that this gave them protection from lions.[5]

These communities have lived with lions for thousands of years, although some have forgotten the potential dangers, or become complacent. Our role as the Niassa Carnivore Project is to remind the younger generation of what's at stake, in an effort to forestall incidents that could have dire consequences for both humans and lions.

In 2008, we camped in the fields of Mbamba Village during the wet season to find out what it is like to have to protect your crops in a wildlife area in the pouring rain, with regular baboon and elephant raids. At that time, only 7% of farmers had a *sanja* or safe house on stilts, and 13% had a secure house on the ground with walls and a door. Nearly 40% were simply sleeping in the open under a thatch roof. People seemed unperturbed about walking alone after dark, and very few of the individuals stationed in the fields at night to chase animals such as bushpigs had any form of protection or lighting other than fire.

We found an elder sitting quietly outside a secure log house that must have taken a huge amount of effort to build. When we asked him why he went to such great effort, he explained that it was necessary to keep him safe from lions. However, while more than 60% of the people we talked to had seen or heard lions in the fields that season, many said they had no time to build such a safe shelter. Building materials such as bamboo could only be sourced far away and were difficult to transport, so they were prepared to take the risk.

Three of the elders spoke of the spate of lion attacks in the 1980s when seven people were killed here, but most made no mention of it. What we learned from these people fed into our ongoing 'Protect yourself with safe behaviours' campaign to reduce lion attacks. People needed to be reminded of the potential dangers, even as memories of significant lion attacks in the past faded. To disseminate our message, we used community radio, door-to-door outreach, posters, lessons and games for children; and we gave people like Mama Amisse a platform from which to share her story: the same message told a thousand ways, over many years and at every opportunity.

In 2021, we redid the survey of the types of shelters people were using in their fields and found 98% of the farmers in Mbamba were sleeping in safe shelters, mostly in *sanja* on stilts, compared to the previous 7%. No one has been attacked by lions or hyenas in Mbamba for many years.

Lions were not seen as an enemy. Our team asked the leaders what would happen if all the lions were gone. An elder answered: 'Our children would go back to being illiterate.' That was a nod to 10 years of a community conservation partnership with that village. Lions and other wildlife had brought them secondary school scholarships, a community tourism camp, significant revenue, 18,000 seasonal worker days in conservation services (building roads, firebreaks and various other construction projects) and money into a community fund that had completed a new maternity unit.

We have seen an increase in the number of lion prides in this area from two to seven, two of which regularly move around the fields inside the community development zone, extremely close to people. This is a precarious peace aided by the fact that there are no cattle here, and by the increasing natural prey like buffalo, warthog, waterbuck and bushpigs. And people are living in safe shelters.

Peacebuilding

These villages offer lessons in coexistence and peacebuilding. Four traditional chiefs live in Mbamba village and have been harmonising the competing desires and needs of their multicultural people for decades. Most people speak three or four languages and families practise both Muslim and traditional religions. There are no sharp separations.

It's undoubtedly frightening to live among lions and always has been. In 2022, our rapid reaction team, with reserve authorisation, shot a young male lion in the reserve that had killed five people, including a young girl, in a year. Our outreach on safe behaviours has not consistently reached all the villages in the west of this vast landscape and, in an area where there have also been no clear benefits or revenue sharing from conservation, there remains a lot of work still to do. Bushmeat snaring is extensive in many areas, reducing the lion's natural prey. There is no recompense for the loss of life.

You cannot buy coexistence. Revenue sharing and related benefits will never be enough to stop retaliatory killing if a person is attacked. Neither should it be. The only reasonable and ethical response is to kill the lions that do it. It's not compensation, but it is justice and does stop the immediate problem. Yet, the public, far removed from the reality of living with lions, is quick to judge. 'Lions were there first,' they say, 'it's a problem with people, not lions.' In Niassa, this is simply not the case and not how coexistence works. Coexistence is not peaceful nor perfectly balanced. Sometimes the costs are too high, and retribution is inevitable.

The supernatural

There is another side to coexistence, which, as biologists, we find more difficult to understand – but it's critical for conservation efforts. In 2002, when we first arrived in Niassa Reserve, Paulo Israel – a historian and social anthropologist – was conducting fieldwork on Makonde *mapiko* dances in Cabo Delgado province in the rural district of Muidumbe in northern Mozambique. It's here that the so-called 'War of the Lions' began when lions started to eat people. Over a six-month period between 2002 and 2003, more than 50 people were killed and many more were injured.

While the 'outbreak' of lion attacks showed many similarities to what was happening in southern Tanzania around the same time, the attacks in Muidumbe were widely seen through the lens of the occult and as more of a social uprising. Lions were thought of as being fabricated through sorcery, and rumours spread about a secret society of 'lion-men' involved in the trafficking of organs harvested by these ghost lions.[6] By the end of this period, more than 24 people had been lynched and more were threatened and ostracised on suspicion of being connected to the fabrication of the lions.

This changed our thinking. Did we ask the right questions in our survey? We realised our questions on lion attacks were incomplete. Coexistence is not just ecology but also life, belief, culture and politics. This opened up our questioning to include questions about 'spirit lions' as being distinctly different to 'bush lions'.

In 1969, in Negomano, a man was said to have been injured by a 'spirit lion' in retaliation for sleeping with another man's wife. In July 2005, an elderly man who was dreaming of being attacked by lions was set upon and injured as he slept outside in his field with his wife. The lions were said to have been 'sent by Chemambo' (an important sacred site).

It's about coexistence

In a wilderness like Niassa, there are no arguments that animals were here first or that this land belongs only to people. Nobody who lives here believes animals must stay in their place and people somewhere else. There are many magnificent moments of coexistence happening every day that get overlooked in our blinkered approach to conservation. A man celebrates the presence of a pride of seven lions in his fields hunting bushpigs and therefore protecting his crops. A lioness gives birth in a small rocky cave right next to the fields, and is celebrated.

Above: Mozambique has become a hotspot for the illegal lion body-part trade. To fight this rising threat and stabilise the lion population, a Lion Coalition project was initiated, which includes collaring lions and tracking the real-time movements of prides.

Left: Footprints indicate that people and lions are living alongside one another.

Here lies the most important lesson we have learnt about coexistence in the past 20 years. Coexistence here does not equate simply to peace; it's dynamic and relies on a shared history with, and understanding of, how lions intersect with society; how people live, how they stay safe, how they rationalise attacks and how lions are part of culture. That does not mean that reducing conflict is not important, with the help of cattle bomas, goat corrals, safe shelters, flashing lights, revenue sharing, performance payments, and substantial amounts of revenue. However, it is simply not sufficient. You cannot buy coexistence.

Taking into consideration the voices, stories, fables, beliefs, opinions and decisions of people who have lived – and continue to live – with lions is critical. If this link between lions, the land, society and culture is broken, coexistence will become impossible; conservation across these vast landscapes where lions and people live together will ultimately fail and will become no more than superficial tolerance.

Coexistence requires a deep-seated belief that nature is inclusive of people; that both wildlife and people matter equally and have a place. It's too hard to coexist, with all the costs involved, without already having a connection with the land and believing you are part of nature. This is why so much urban and farmer conflict with wildlife is so difficult to resolve and frequently unsuccessful.

As our journey in Niassa continues, we are left with the feeling that coexistence with lions is much like happiness: fragile, transient, you know it when you see it but then it is gone. It is not a steady state. Helen Garner in her recent 2023 essay on happiness asked: 'What is happiness, anyway? Does anybody know? It's taken me 80 years to figure out that it's not a tranquil, sunlit realm at the top of the ladder you've spent your whole life hauling yourself up, rung by rung. It's more like the thing that Christians call grace: you can't earn it, you can't strive for it, it's not a reward for virtue. It exists all right, it will be given to you, but it's fluid, it's evasive, it's out of reach. It's something you glimpse in the corner of your eye until one day you're up to your neck in it. And before you've had time to take a big gasp and name it, it's gone.'

For us, this is a perfect framing for coexistence: evasive, dynamic and fragile and built into the human condition through a shared history.

'For lions to coexist with people, it is not so much about protecting people from lions or lions from people, but rather safeguarding the connection between people and nature and appreciating our shared evolution on this planet.'

Above: Mbamba Village is located inside the Niassa Special Reserve in a remote, wildlife-rich area close to the Lugenda River in Mozambique. The nearest town with a hospital or a bank is 75km away. The village was settled in waves of clan movements spanning precolonial times and today most of Mbamba's 2,000 inhabitants are still engaged in subsistence agriculture with a strong reliance on fishing. These days, the residents also earn some income through work in conservation and small amounts of tourism.

Left: Lions are a powerful symbol of strength, courage and nobility across the world, including in Niassa Reserve. Lions feature strongly in art, songs and theatre of the people who live inside the reserve. These grass lion masks were made by a local artisan group, Kushirika, for a lion festival that is held each year in Mbamba Village. The people of Mbamba Village live closely with lions and encounter them frequently at night, when the lions enter their fields to hunt crop-raiding bushpigs.

Lugenda River, Niassa Special Reserve, Mozambique. © Dr Colleen Begg

MOZAMBIQUE

31

Gorongosa's lions

A dark-maned, retro-looking African lion is the symbol of Gorongosa National Park in Mozambique. Revered and feared in equal measure, lions are an important animal in local culture. The Chitengo family (descendants of Chief Chitengo who gave the park's headquarters its name) have the lion as their totem animal.

James Byrne

Before Mozambique's long and tragic civil war from 1977 to 1992, Gorongosa had a healthy lion population of about 200. Old film footage from the 1960s and 1970s shows them lying around in the shade – carefree, regal and fat – surrounded by huge herds of buffalo, zebra, and wildebeest.

But during and after the war, almost all the animals disappeared – killed for their meat by hungry soldiers or by trophy and bushmeat hunters exploiting post-war lawlessness. With very little prey left to hunt, the lion population crashed to 10 or even fewer individuals.

In 2008, the government of Mozambique and the US-based Gregory C Carr Foundation signed a 20-year 'public–private partnership' to restore Gorongosa National Park. They called their agreement the Gorongosa Restoration Project. The agreement was extended in 2018 and will continue until 2043 – a total of 35 years.

The restoration team hired, trained and equipped rangers to protect the park's 405,000ha. Select animal introductions were made to boost the remaining nuclei of different wildlife species, including 210 buffalo and 180 blue wildebeest.

At the same time, the park began extensive social development programmes in health, education and agriculture to support the livelihoods of local communities.

During their daily patrols, rangers began removing thousands of wire snares and gin traps. These had been set over many years by illegal networks to catch animals like warthog or impala for the bushmeat trade in the cities. Although lions were not the intended target, they could also get injured or killed if they were unlucky enough to step into a concealed snare. This was slowing the pace of lion population regrowth in the park.

By fitting some lions with GPS collars, scientists were able to monitor and track most of the lion prides. If a lion got snared, the GPS signal would go static, indicating a potential problem. Using the GPS position, vets could find the lion, tranquillise it, remove the snare, and treat the wound. Many snared lions survived and prospered – demonstrating the extraordinary strength and resilience of the species.

After most of the snares had been removed and rangers had created a safe park thanks to their daily patrols, prey populations boomed. In just 10 years, the park's large-animal population grew from 10,000 to over 100,000. With plenty of prey to hunt, lions roared back. Today, Gorongosa National Park has at least 230 lions, exceeding the pre-war population. The population continues to grow and the park is well on its way to becoming a lion stronghold and a critical sanctuary for the beleaguered species.

Gorongosa also provides hope for the future of other endangered or vulnerable large carnivores. Once the lion population had grown back to full strength, the restoration team began reintroducing other members of the carnivore guild: African wild dogs, leopards, spotted hyenas and jackals. Today, all these species are thriving again, bringing back ecological balance and heart-stirring wildness to a place once considered a lost cause.

Gorongosa, Mozambique. © Brett Kuxhausen

Gorongosa National Park, Mozambique. © Charlie Hamilton James

Gorongosa National Park, Mozambique.
© Charlie Hamilton James

Hwange National Park, Zimbabwe. © Mike Myers

The lions of Zimbabwe

PORTFOLIO

Makalolo, Hwange National Park, Zimbabwe.
© Diana Sutter

Chilojo Cliffs, Gonarezhou National Park, Zimbabwe.
© Anoet du Plessis

Makalolo, Hwange National Park, Zimbabwe.
© Diana Sutter

Makalolo, Hwange National Park, Zimbabwe. © Diana Sutter

Ruckomechi region, Mana Pools, Zimbabwe. © Mike Myers

ZIMBABWE

32

Blood under the carpet

Lions are killing livestock outside the borders of Hwange National Park and there is a link to trophy hunting, but that information is being kept under wraps.

Brent Stapelkamp

A worrying trend is happening among lions in Hwange National Park: subadults and females, some pregnant, are leaving the protected area and raiding cattle. This is uncommon behaviour and needs to be investigated.

Unlike hyenas, which can live among people, these lions tend to raid and leave. While adult males usually avoid people at all costs, around Hwange, the most common raiders are subadult males (between two and four years old, but occasionally younger) and adult females, either pregnant or with dependent young. There's something wrong with this picture, which got me asking the question: Why these specific demographics?

I have heard it said that lions prey on livestock because they are easier to catch, have more fat and even (I love this one), taste saltier. I might agree with the first point comparing the 'fight or flight' of a buffalo to a Nguni cow, but for the rest, I'm not convinced. If livestock were the preferred prey, then surely it would be the adult males, the largest and strongest lions, that would be out there taking advantage of such a food source. Yet we never see them around settlements.

Studies have found that, when given a choice, lions will opt for natural prey over livestock. So, what else could it be? Perhaps the price of cattle raiding is too high and the risks too great for the largest animals?

Why then are lionesses and subadults willing to pay the exorbitant cost? Why does a lioness that has never left the protected area suddenly take her cubs into the communal lands and start eating goats and donkeys? Why, when the lion guardians track her down and 'haze' her with vuvuzelas and fireworks, does she return night after night? Does she like the smell of gunpowder in the morning?

Clearly, there is a story that isn't being told in lion-conservation circles. In some cases, perhaps the solution to this conundrum has not been found. But in other cases, it's more calculated and sinister and being swept under the rug. I am taking this opportunity to whistle-blow.

We all know about infanticide in lions, the killing of cubs usually by new male lions when they take over a pride. Male tenure over prides and their territories tends to be relatively short, perhaps three or four years, if they are lucky. When a lion

Cecil was arguably Hwange's most famous lion. He was large and handsome, and he controlled most of the territory around southern Hwange's famous plains. Cecil was shot with a crossbow, many say illegally, as he was entering his prime, and was left to die an agonising death.

Blood under the carpet ~ **297**

takes over a pride after displacing the incumbent male, there are often cubs and subadults in the pride who are still dependent on their mothers. These lionesses will not mate with the new male. ('Still dependent' refers not just to those cubs still suckling; if the mothers and the pride still hunt for the youngsters, they are considered dependent.) The new pride male cannot afford to wait, so he kills the cubs and subs, after which the lionesses come into oestrous in a matter of weeks and are receptive to his attentions.

In nature, replacing existing cubs with genetically stronger progeny makes perfect sense. When a lion attacks, beats and replaces the incumbent pride male, the newcomer probably has some genetic advantage that merits being put into the system.

A colleague, Dr Nic Elliot, researched dispersal-aged lions in Hwange for his doctoral thesis.[1] In his extensive study he established that, where a pride male held tenure for a long period, his male offspring left the pride between 3.5 and 4 years of age. Where he had lost tenure of his pride or been displaced, his offspring might leave their pride at as young an age as 13 months old (in an extreme case). There is a huge difference in experience between a four-year-old male lion and a 13-month-old one. The four-year-old weighs about three times as much as the younger one and will have participated in many hunts. The youngster still has milk teeth and cannot even roar. Dr Elliot noted that if a disperser left his natal pride before 31 months, his chances of mortality increased. He called it 'delayed infanticide'.

We know, from research, that lionesses prefer male lions with larger manes. They also have a distinct preference for the darker ones over their blonde counterparts and often give them preferential mating rights. Trophy hunters also prefer heavier-maned lions and fall over themselves for the darker ones. Cecil the lion, probably the most famous lion ever, was shot by a hunter as soon as he arrived at the bait, despite his blonde coalition partner feeding in front of the hunters for an hour. That is because Cecil was black-maned.

So, chances are that the largest, darkest lion in a landscape is a territorial male. He is fitter and stronger than the others and, after receiving preferential mating rights, he is the father of the pride's cubs. He's the one that ends up in the hunters' sights: baited and shot. The happy hunter takes his trophy home knowing he has done a 'good job' for conservation, or so he has been told.

What he does not see and is not told about is that two or three weeks later, the absence of his trophy lion has been noticed. His silence at night and the stale scent markings proclaim that his land is vacant, and former competitors start closing in. The first time the pride females realise there are new males on the scene is usually when they're feeding on fresh kill, with vultures spiralling overhead. Suddenly there is chaos – snarling and growling, with cubs flying in every direction like tawny popcorn. The new contenders for the top spot have arrived, and what follows is not pretty at all.

Infanticide on paper sounds nice and clean. But, like every mother on Earth, a lioness' strongest instinct is to protect her offspring and it is not clean at all. We have seen lionesses steal away with all the cubs and run for the hills while their sisters, cousins or aunts fake oestrus and distract the new males with sex (you can tell this is happening because *he*, not she, initiates the actual mating.) It sometimes buys time for the cubs to get to safety. Occasionally, a lioness will fight to protect her cubs and subadults; but in Hwange, as in many other healthy lion landscapes, she will be taking on a male coalition commonly consisting of three or four lions, and there is a high chance she will die or be severely injured. So, what does she do?

She flees with the cubs. They run and run and don't stop until the threat of the male lions is gone. And where is it, in modern Africa, where the male lions will not follow? That's right: out of the protected areas and into the human settlements.

This is why we see lionesses with cubs and subadult males raiding communal lands. That's why, when we chase them back to the park, they come out again and again. It is not so much that they're attracted to the communal lands and livestock, but that they are forced there. Once among humans, they struggle to find natural prey and turn to livestock, which is generally poorly attended. And they can do severe damage. A lioness called Frisky, which had never before killed livestock, was forced into this scenario when her pride male was struck down. She killed 35 goats in the ensuing weeks before she and her cubs were also taken out.

If we wait endlessly for science and peer review before we act to stop lion hunting, we may lose our most charismatic species and cause a lot more pain and suffering for people who live with lions.

Of course, the loss of a pride male can occur naturally, and it does, but this is rare. I perhaps saw only two or three pride males die naturally in a decade of lion research. They can die on train tracks or in snares, but as Dr Andrew Loveridge found and published in 2016,[2] trophy hunting was the largest source of mortality for male lions in Hwange. Some lion conservationists and people who are pro-trophy hunting respond by saying trophy hunting is not the major threat to lions; that habitat loss and conflict with people over livestock and the subsequent retaliatory and pre-emptive killing is a far greater threat. I disagree: trophy hunting may just be the major driver of lions leaving the protected areas and conflicting with communities.

Loveridge noted that trophy hunting of adult males was the largest source of mortality for that demographic. Let me add that it is also the largest source of mortality for adult females, their cubs and subadults. I remember seeing that something like 40% of lion mortality across Africa was as a result of conflict and only 10% from hunting. Well then, added together, this is 50%! And all driven by trophy hunting. The rest of the mortality comes from natural causes, collisions with trains, etc. Here's the reasoning: people argue because trophy hunting is only 10% and conflict is 40% then let's forget hunting and concentrate on the conflict. I counter that we know from Hwange that hunting drives the conflict.

Cattle raiding is worst in the southeast of Hwange around Ngamo, a large flatland that fills with shallow water each year and looks like a small Okavango delta. It is home to a local population of wildebeest and zebra that migrate in and out, as well as large herds of buffalo. Water is pumped in during the dry season. This creates a perfect territory for lions and, indeed, they have always been there. The problem is that, on its northern boundary, there is a railway line and on the other side is a hunting area. The boundary is an old veterinary fence and beyond that is a communal land.

Each year, the pride male or males are being attracted into the hunting area with bait, and shot. If one of a coalition is shot, the remaining partners may not be able to hold the territory when a new coalition arrives; they are then chased away, and often end up being shot, too. The new males move in and, as described, the lionesses flee with their progeny, under the vet fence and into the settlements. There they kill livestock and are either shot (often by the same hunters) or snared and speared to death by angry people who have lost livestock.

If the lioness or lionesses survive but their cubs and subs are killed, they return to their territory and mate with the new males. Just as they start to settle down, the next hunting season rolls in and the process starts afresh. Lionesses may have seven litters and not see a single one to adulthood. Meanwhile, upwards of 100 cattle get killed so hunters offer assistance with the 'problem animals' and shoot the lionesses or bring in paying clients to pull the trigger.

I only noticed this pattern because I was investigating the conflict between lions and livestock owners in these hotspots. One of my roles was to respond to and record all conflict reports with GPS location, such as the type of livestock and distance from the park. I had to collar the lions and keep their collars maintained and got a really good feel for what was going on. I mentioned the pattern to my boss at Oxford's Wildlife Conservation Research Institute (WildCRU) and he told me to keep collecting data from all the moving parts.

I went to skinning sheds and collected bloodied lion collars from American and Spanish hunters while they removed the skins from lions I knew well. I collected dead lion cubs 'killed by dogs' but riddled with spear wounds and sat with angry people who wanted to know why 'Brent's lions' were killing their animals. I caught and collared females and subadult males as well as new males. I predicted which lions on the edges of the territories would be the next 'new males' and collared them, too. We had this story wrapped up tight.

'Wow,' I thought, 'what a significant paper this will be for lion conservation and conflict mitigation.' I spent weeks away, focusing on data collection, all the time believing it would be worth it. Generally, I was viewed with suspicion. I was the man who was collaring the lions. People collar their cattle with strips of belting and tie bells to them to show ownership. They assumed that because I collared the lions, I owned them.

When we had GPS-collared cattle to see where they overlapped in space and time with collared lions, a collared lion killed a collared cow. People started to say the collars 'talk' to each other and that was how the lion knew where the collared cow was. I was called to a meeting in one village where every attending man arrived armed with an axe. I thought it was firewood day, but I was the intended firewood if the meeting did not go to plan. Desperate to demonstrate how the collars worked, I sent a small boy off to hide a collar for me to find, and I tracked it down, together with an entourage of armed men.

What made my efforts more difficult was that trophy humters were threatened by lion research and feared what it might unearth. They would stir the pot and rush in to help the stricken community with its 'lion problem'.

I collected data for more than two years. In 2015, a lion called Bush was killed in the Ngamo Forest. He was a pride male and his death left Bhubezi, his brother and coalition partner, alone. Bhubezi was chased out of the territory and, just up the tracks, another lion was lured and killed in front of his coalition partner. This was Cecil and it nearly broke the internet. The world's attention came to bear on the lions of Hwange. It was the perfect opportunity to speak up for lions and would have been an opportunity to publish this crucial study exposing the impact of trophy hunting pride males and of lion conflict with communities.

The Cecil story was extremely stressful for my family. I was at the epicentre and it took its toll. We had also been developing ideas about how to tackle human–wildlife conflict more holistically. One thing led to another and I resigned from the lion research project. My wife and I then started a community-based trust.

I remained a member of the African Lion Working Group and met with and mixed with my peers. Through that forum, I was shown a report that WildCRU had written for the British government on the impact of trophy hunting on lions and lion conservation.[3] To my amazement, there was no mention of our findings or our study, nor that it was in preparation. It was central to the issue, why was it not in there? I knew they had the data.

I replied to the group that I found the report to be hunting-biased and not the gold-standard scientific report we would expect from Oxford. I was met with a mixed bag of responses from my lion working group peers. Many scientists and conservationists said that, since I'd mentioned it, they were seeing similar links between a lion being hunted and a spike in conflict in their areas.

I was told by WildCRU that, because I was no longer one of the team, I was out of the loop and the paper was awaiting peer review and would be published 'soon'. I enquired why it was not mentioned as in preparation if it was awaiting review. No answer was forthcoming and so I waited. That was November 2016!

In April 2017, a report was published by WildCRU on the gaps in knowledge around lions and trophy hunting and there, for the first time, the paper was mentioned.[4]

I wish people had said I was anti-anthropogenic mortality and not anti-hunting. I'd be invited to more braais!

Six years later and many lions have since died. Many people have lost their only source of wealth (their livestock) and lions have been increasingly portrayed as 'crazed' killers out to destroy lives. That unfairly tarnished image is used politically to push for more problem-animal control, wildlife trade and consumptive use.

In the absence of a timeous publication into research on the disastrous impact of trophy hunting – on lions and surrounding communities – I offer this information to you as a whistle-blower. If we wait endlessly for science and peer review before we act to stop lion hunting, we may lose our most charismatic species and cause a lot more pain and suffering for people who live with lions.

The Victoria Falls to Bulawayo railway line often serves as the only barrier between the safe sanctuary of Hwange National Park and the adjacent hunting concessions to the east. There is no buffer zone between. For a wandering lion, the distance between life and death can be just a few metres.

Desert-adapted lion from the Hoanib Valley, Namibia. © Dan Achber

The lions of Namibia

PORTFOLIO

Etosha National Park, Namibia. © Inki Mandt

Desert-adapted lions in northwestern Namibia. © Inki Mandt

Etosha Pan, Namibia. © Hamman Prinsloo

Huab River, northwestern Namibia.
© Inki Mandt

NAMIBIA

33

The desert lions of Kunene

Poor planning has radically reduced the survivability of Namibia's desert-adapted lions. There are solutions, but they need to be implemented now.

Izak Smit & Ingrid Mandt

Since 2011, my wife, Inki Mandt, and I have spent about three weeks out of every two months between the Ugab and the Hoarusib rivers in the Kunene region. We were originally lured into the oldest desert in the world by photography, but the desert-adapted lions changed everything. Our interest became something far deeper and more captivating.

What are desert lions? The 150mm isohyet not only serves to connect places of equal rainfall but is also used to distinguish these desert-adapted, free-roaming lions from the others. Those to the east of it are considered to be 'normal' Etosha lions, while those below the Etendeka Plateau to the west are considered to be desert adapted. These adaptations include a lighter, more athletic build, far less dependency on water, smaller prides, thicker pelts (particularly near the coast), the ability to cover great distances, finely honed hunting skills and wider prey diversity, including marine animals.

In the late 1980s and early 1990s, we regularly came across fresh lion tracks in areas from the Ugab River in the south right up to the Nadas River in the north, which challenged the claim that there were far fewer lions in the area back then. Although we never saw these elusive desert phantoms outside of the Palmwag concession area, they were clearly there.

Our interest changed dramatically when I spent over two years as a chopper pilot on contract in the Busanga Plains in Zambia. I assisted the local lion researcher and a film team shooting the documentary, *Swamp Lions*. We also started reading publications of the Desert Lion Project (now the Desert Lion Conservation Trust) and met its researcher, Dr Philip Stander, in the Huab River while he was monitoring two lionesses denning with eight very young cubs.

Our first glimpse of the desert phantoms were the lionesses, XPL 75, or Angela, and her sister, XPL 76, against the fantastic backdrop of the Huab valley in 2012 (X indicates the region, Xhorixas, PL for *Panthera leo* and the number is the order in which they were collared by the researcher). We were smitten by the uniqueness of these extraordinary ghosts, and geared up to capture quality images. Our visits became more regular and longer. After the good rains from about 2008 through 2011 in this region, game was plentiful, and the otherwise inhospitable environment went into a boom cycle. Our red Land Cruiser camper became a familiar sight and soon we were known by the local communities and farmers as the Red Cruiser Lion People.

Initially, the odds against finding the lions were as high as winning the Lotto. Yet, gradually and painstakingly, we got to know most of the desert-adapted prides roaming outside the protected Palmwag concession. The first was Angela's Pride in the Huab valley. It had become so successful that a splinter group migrated to populate the Ugab River valley in late 2013. On the back of good rainy seasons over the previous five years, they flourished. At one stage, between the Huab and Ugab rivers, there were 20 lions. They would typically split up after reaching more than about five in a pride, dispersing to new territories as part of their adaptation to the relatively low prey density of the desert.

The best part of our time there – and probably of the desert lions – was when human–lion conflict (HWC) was relatively low because of favourable prey density, keeping the prides to their natural home ranges. Conflict with humans was then restricted to the dispersal of subadults from their home ranges, an essential process to ensure genetic diversity. Being inexperienced and clumsy apprentice hunters, these youngsters would sometimes seek easy prey in farming lands skirting the wildlife and tourism areas.

It was a time to rejoice, with this arid Eden showing off its splendour. Fauna and flora big and small graced the otherwise hostile landscape and natural springs yielded water in abundance. We saw wildflowers we had never encountered before or since. Plains like Giribes, Otjihaa and Klein Serengeti were teeming with oryx, zebra and springbok and in the ephemeral rivers were kudu and smaller antelope. Birds of prey thrived and our hearts soared with them. The abundance of smaller predators like jackal, Cape fox, aardwolf, bat-eared fox, honey badgers, and brown and spotted hyenas indicated a healthy ecosystem and the young of all species were common.

In time, through our careful approach and strict protocols, the lions got to recognise our vehicle and we were privileged to enjoy their trust. We would spend days with them, often losing the feeling in our legs after sitting dead quiet for hours clicking away at the superb images they provided. A few times we got treated by denning females parading their barely mobile cubs and allowing them to play near our vehicle. There were many emotional episodes that caused us to unashamedly shed a tear out of appreciation and understanding of their vulnerability.

Often, with shade being so scarce in the desert, the lions would rest in the shadow of our parked vehicle. We never interacted with them, and would always whisper to minimise any desensitisation that could compromise them. We would take great care when driving the lions away from livestock and kraals in conflict situations, not wanting them to associate us and our vehicle with the bad experience of the chase-offs.

The downturn

If only we had been prepared for the emotional roller-coaster ride awaiting us. Many of the cubs that we had known from birth were being shot or poisoned as subadults, while still in their prime. In time, as the prey density fell to alarmingly low levels as a result of over-hunting and natural climatic variation, the lions were forced to prey on livestock and even chickens and dogs in order to survive. The formerly euphoric picture was about to turn very ugly. We saw this coming and, despite proposing sound, workable solutions to the relevant ministry and associated NGOs – and strongly advocating for their implementation – we witnessed this horror playing out in front of our eyes. Bureaucratic stonewalling and ineptitude rendered us hopeless and helpless.

Not long after the times of abundance, during which we noticed a huge increase in hunting, a severe drought descended. It became known that we were familiar with the lions and their movements, and we were approached by the Torra Conservancy community elders and committee to help mitigate the increasing human–predator conflict following periodic raids by the desert lions. The desperate farmers had planned to use the Huab valley for emergency grazing, knowing only too well that the Huab Pride, consisting of about nine lions at this time, roamed the area as part of their home range.

Namib Desert, Namibia. © Inki Mandt

In time, through our careful approach and strict protocols, the lions got to recognise our vehicle and we were privileged to enjoy their trust.

Namib Desert, Namibia. © Inki Mandt

The desert lions of Kunene

Seeking answers

We volunteered to assist and a communal kraal was erected to the east of Peter's Pools in the Huab valley. We provided motion-triggered floodlights as well as fireworks, torches and airhorns to deter nightly predation. We spent many nights there. Our interventions worked well, resulting in the absence of lions, and only a few livestock killed over a long period – all animals that had spent nights outside of the kraal. It resulted in a good relationship with the farming community and conservancy.

We were sucked ever deeper into this protection and established a voluntary association, Desert Lions Human Relations Aid (DeLHRA). We conducted a pilot project providing shade cloth to cover their dilapidated, rudimentary kraals, mostly built from sticks and mopane branches, as we found lions avoided these. It became clear lions would not jump over obstacles without knowing or seeing the landing spot beforehand. They lack the agility of leopards, which can pause at the top of the obstacle and jump back to safety if the landing spot poses a risk. Also, by removing the visual cues on livestock inside the enclosures, lions seemed less interested and would not even bother to probe the barrier between them and potential food, no matter how flimsy.

Word of this successful strategy soon spread and we were inundated by requests for help in all the areas where the lions were active. This then led to a memorandum of understanding between DeLHRA and five conservancies in which the desert lions roamed.

Despite scepticism and attempts to discredit us, after dishing out many kilometres of shade cloth to conservancies, a success rate of 100% was achieved and the night raids by desert-adapted lions on the kraals were halted in their tracks.

The beginning of the end

From 2014, we noticed the larger prey of these lions was fast disappearing, and annual game counts, along with our own, corroborated this. Upon enquiring, we learnt from the NGO conducting and co-ordinating the game counts that 'sustainable' hunting quotas – allocated to the conservancies in terms of

Huab Valley, Namibia. © Inki Mandt

a community-based natural resource management model – turned out to have been over-optimistic, inflated and erroneous, owing to a flawed 'scientific extrapolation formula' applied for these purposes.

It was now 2015, the drought had started to bite and criticism of overhunting of desert-dwelling game was pouring in. It came from seasoned former nature conservation employees, tour guides, lodge owners, locals and tourists. An article in *The Namibian* newspaper by Christiaan Bakkes, a respected conservationist and guide with decades of hands-on experience, ran under the headline 'End of the Game' and his book *Plunderwoestyn* (Plundered Desert) ruffled many feathers. This led to his being sanctioned by the Ministry of Environment, Forestry and Tourism and associated NGOs to the point where he had to leave the country to find employment. Research and work permits are still handy tools to keep those dependent on them in line.

The disappearance of prey species was by now clearly congruent with the sharp increase in human–lion conflict around livestock predation. Other NGOs started building large, expensive, sturdy communal kraals that were, surprisingly, largely unused. When we asked tribal chiefs and farmers why, we learnt that they would not kraal their livestock together with another farmer's as this led to conflict. Chief Nicolaas of Mbakondja asked us: 'If your pregnant ewe and mine go into the same kraal tonight and tomorrow we find a new lamb in the kraal, to whom does it belong?'

Farmers insisted on having their kraals near to their houses where their dogs could warn them of predators. This was their custom and they would not be persuaded to change. We respected this, always asking how they wished us to help. We supplied the material and they, as the owners, had to roll up their sleeves to fortify and improve their kraals – an arrangement that worked well.

Losing the war

Night-time attacks – the way lions usually and easily took livestock – became something of the past, as lions became desperate. They adapted by ambushing livestock on their way to the kraals for the night or coming out of the kraals to graze in the morning. Herdsmen quit their jobs because of the danger posed by the predatory lions, leaving many farmers in trouble.

We appealed to the Ministry of Environment, Forestry and Tourism, associated NGOs and the Desert Lion Conservation Trust to buy into, incorporate or allow us to implement our plan to establish a stakeholders' network. This would act as a multipronged rapid-response unit to pre-empt daytime attacks using satellite-collar positions of lions breaching their geo-fencing. The rationale was that stakeholders benefiting from the desert lions, such as lodge owners, should become involved in mitigating human–lion conflict as they had trained guides, infrastructure and transport, and were strategically located near the conflict hot spots, enabling rapid response times. The affected communities would provide the workforce and lodge owners, in turn, would depend on the wildlife for tourism. This plan was snubbed – without discussion – by the Northwest Lion Working Group and the Ministry of Environment, Forestry and Tourism.

What had gradually been put in place came unstuck. An NGO was provided with three sponsored rapid-response vehicles, which patrolled the vast area between Opuwo in the north and the Ugab in the south. Not long after, a couple of these vehicles were lost in a flash flood in the Hoanib River. Other measures – such as employing over 40 salaried lion rangers without transport, tracking equipment or remote communication, and erecting early-warning towers near some villages, at great cost – did not change much. Statistics on lions killed in conflict and subsequent retaliation confirmed this. Shooting and poisoning of lions increased as responses invariably became reactive instead of pre-emptive or proactive. Farmers tend to take matters into their own hands when they find promises of help are ineffective, and either do not arrive or arrive after the fact.

Another problem arose: increased conflict with hungry lions led to the false assumption that there were simply too many lions. However, videos and photos of starving lions, some collapsing next to kraals during the day, underscored the fact that the lions had left their prey-less home ranges to seek food at farmsteads for survival. Game counts corroborated this.

Dogs, chickens and anything else except humans became targeted by lions. At the Save the Rhino Trust (SRT) camp in the Ugab River, the caretaker, Johannes, was trapped in his hut for three nights when the otherwise shy and well-behaved collared male lion Doros (XPL 115) came back night after night to catch what was left of his dogs and chickens. Not surprisingly, he left his job in a hurry – but returned after the lions of the Ugab were exterminated. The Ministry of Environment had to euthanise one of the starving Huab lionesses (XPL 75); some starving individuals were shot at Opdraende farm for preying on livestock; and the rest (four) had to be translocated to a sanctuary after we found them grossly emaciated and in an advanced stage of starvation. The Huab Pride was no more and the Ugab Pride had been reduced to one female, which by now may have starved to death as well.

Outside of the protected Palmwag area and the Skeleton Coast, where some of the lions also succumbed to starvation, perhaps just one desert-adapted lion may still be alive. Considering that the Huab and Ugab prides at one stage consisted of about 20 lions, this is a tragedy.

There are solutions

The saddest part of this story is that solutions do exist. If implemented, however, they would undermine the incomes of many sponsored and funded NGOs that depend on the continuation of the problem. Sound, workable solutions have been stonewalled as they might upset cosy arrangements and agendas benefiting some.

We can fix this. We need to identify unoccupied land that's unfit for farming but ideal for tourism and wildlife, and have it proclaimed as a protected area. The land from the Ugab River up to the Springbok River near Wêrelds End – the natural habitat of desert-adapted elephants, lions, endemic black rhino, giraffe, oryx, kudu, zebra, hyena, aardwolf and many other fauna – presently generates no revenue. It's rapidly turning into a wasteland, devoid of wildlife through abuse. It has the potential, like the Palmwag concession area, to generate substantial income and job opportunities for the communities, not to mention the benefits it could have for conservation.

The subconcession holders in Palmwag offer exclusivity to their guests at a high premium and, as a consequence, thrive financially. This system needs to be copied and pasted into the Ugab and Huab areas where it would create skills, jobs and strategic business partnership opportunities for the communities. On a small scale, fodder such as barley could be grown hydroponically in old shipping containers, and could be harvested within seven days, on a rotation basis, by subsistence farmers. In hotspots where possible threats from predators exist, livestock could then remain safely in their kraals while being fed. They would expend less condition-zapping energy having to walk great distances to find the sparse grazing, and the crippling effect of the cyclic droughts would be less pronounced.

Staple foods such as higher-value fruit and vegetables could be hydro/aquaponically produced and sold to local lodges. Any extra fresh produce could be transported to marketplaces in the refrigerated trucks servicing lodges and camps in these areas, which usually return empty. Community projects in areas suitable for solar farming and near the electrical grid would also be a possibility.

The sounds of silence

The current status quo reflects poorly on the much-revered community-based natural resource management (CBNRM) model as the self-perpetuating effect leading to the degradation of this sensitive ecology deepens. At the current rate, in our lifetime we may well hear the last roar of a lion in the oldest desert in the world.

We have already heard the last roars of the Hoarusib Pride near Purros and the Okonkwe Pride south of Purros (all killed in human–lion conflict) as well as the Huab Pride south of Wêrelds End and Bergsig, and the Ugab Pride north of the Brandberg (some killed in HLC, some translocated, some confirmed to have died of starvation). Only remnants of some prides in the Palmwag concession area remain, including the Hoanib River Pride. In the near future, with the lack of genetic diversity brought about by an isolated, less-than-viable population coming into play, we may soon see the last desert-adapted lions in this area.

In the near future, with the lack of genetic diversity brought about by an isolated, less-than-viable population, we may soon see the last desert-adapted lions in this area.

Lions in the early morning mist, Skeleton Coast, Namibia. © Peter and Beverly Pickford

Hoanib Valley, northwestern Namibia. © Inki Mandt

Okavango Delta, Botswana. © Ona Basimane

The lions of Botswana

PORTFOLIO

Okavango Delta, Botswana. © Ona Basimane

Okavango Delta, Botswana. © Dana Allen

Okavango Delta, Botswana. © Dana Allen

Okavango Delta, Botswana. © Ona Basimane

Okavango Delta, Botswana. © Diana Sutter

Okavango Delta, Botswana. © Garth Thompson

Xigera lions, Okavango Delta, Botswana. © Mike Myers

Chobe River, Botswana. © Margot Raggett

Savute region, Chobe National Park, Botswana. © Martin Harvey

Central Kalahari Game Reserve, Botswana. © Charlie Lynham

Two much younger and smaller lions (left) challenge, injure and drive away the older, larger and stronger dominant male in the Mokolwane Concession, Okavango Delta, Botswana. © James Ramsay

Okavango Delta, Botswana. © Michael Poliza

Northern Okavango, Botswana. © Christian Ungureanu

34

The lion miracle in the Makgadikgadi salt pans

In what was once silent emptiness, planning and care have coaxed part of the Kalahari Desert back to a place of plenty for wildlife, including lions.

Super Sande

Northern Botswana is one of Africa's largest contiguous – and most important – wildlife refuges, stretching from Okavango in the west through the Moremi, Kwando, Linyanti and Chobe to Kasane in the northeast. It's a little over three-and-a-half million hectares of pure wilderness and thriving biodiversity, one of the world's most beloved UNESCO World Heritage Sites, supporting well over 1,000 lions.

This vast wilderness is the heart of the Kavango–Zambezi Transfrontier Park (KAZA) initiative. It staddles large parts of southern Angola and Zambia, as well as much of northern Namibia, Botswana and Zimbabwe. The park is 25% larger than California and much bigger than Germany and Austria combined.

If governments implement the dream, which the presidents of the five KAZA countries have agreed on, it could expand northern Botswana's thriving wilderness tenfold and become one of the most important and largest wildlife areas on the planet. It would also contain the majority of lions in Africa.

The rest of Botswana has smaller groups of lions because the harsh, dry conditions of the Kalahari Desert that make up most of central and southern Botswana cannot support the concentrations of antelope and lions found up north. Only the famed Kalahari desert lions live here. The good news for Botswana is that the Makgadikgadi Pans are changing all this and are now an important new major high-density lion region.

Pacing the change

I was born in the small village of Senyawe, north of Francistown in the extreme east of Botswana, close to the Zimbabwe border. My first job was in the Makgadikgadi salt pans and I still work here today.

In those early days, the Makgadikgadi was a dry, featureless desert. Nothing moved except shimmering mirages. There were no resident lions and if a lion appeared, it would be nomadic, shy and reclusive and quickly move on. If we ever came across their tracks, we'd cover them with rocks so we had something exceptional to show our guests.

Uncharted Africa runs a concession here (another word for a private reserve) of around 364,000ha of flat desert grasslands and salt pans that abut the

Zebra migration, Makgadikgadi Pans, Botswana. © Jose Cortes III

eastern boundary of the Makgadikgadi Pans National Park. That park, too, was waterless, with few animals and no resident lions. If you were lucky, you might have seen the occasional brown hyena or meerkat.

For decades, a few hardy travellers from across the world would book to stay in Jack's Camp, San Camp or Camp Kalahari simply to enjoy the vast open spaces. The emptiness and true wilderness were our only selling points. The sole human resident was Jack Bousfield, a reclusive and fun character who lived at Lazy J, the forerunner to Jack's Camp. There were no telephones or cellphone signal, no electricity pylons. It was pure vast wilderness and, fortunately, still is today.

The Makgadikgadi in those early years was bone dry except for a few weeks of the year in January to late March when seasonal rain filled the depressions and held some water. Short, palatable grasses quickly sprouted and turned green, triggering the arrival of the zebra migration (much like the East African wildebeest migration).

Once our pans dried up and the grasses lost their moisture and nutrients, the zebras left, heading for the Chobe River nearly 300km to the north or the fresh waters of the Okavango Delta to the west. This round trip migration between the Chobe and the Makgadikgadi is still the longest annual zebra migration in the world.

Fire and water

Back then, rampant uncontrolled wildfires caused large swathes of grasslands to burn. These originated from the community areas to the east and were swept westwards through the reserve by the prevailing winds. After each burn there was nothing left. Hardly an insect survived and any nutrients in the ashes were quickly blown away, further depleting the soil. To combat these all-too-regular fires, in the hope of bringing more wildlife back to the area, we created firebreaks and quickly learnt to become highly proficient firefighters.

With fewer fires, the insects slowly started coming back – then the animals that lived off them arrived: bat-eared foxes, aardwolf and anteaters. We have never allowed hunting in our concession, so the wildlife slowly started to trust, learning that humans on game-viewing vehicles did not pose a threat.

The biggest positive change came in the last decade when the park authorities introduced a small number of waterholes into the Makgadikgadi Pans National Park and within our concession. We are fortunate that in parts of the Makgadikgadi there's a 40km belt of pure fresh water lying just below the desert surface. Elephants must have known about this. They migrated in and out of the area and hung around above the water, bringing palm seeds within their stomachs. These germinated and now ilala palms dot the Makgadikgadi in a long meandering belt directly above this underground freshwater lake.

All a borehole driller had to do was to look for the palms and drill to a depth of 9–16m, where they would find ample pure fresh potable water above a band of hard silcrete rock. (If the driller broke through the silcrete, however, this would expose the foulest, smelliest, saltiest water possible, which is useless for human or animal alike.)

Big changes

The combination of fewer fires and eight permanent waterholes changed matters for the good in the Makgadikgadi. Once the zebras discovered the ample fresh water at the new waterholes, they hung about for longer. Elephants also found them and have stayed and thrived on these plains. Today around 1,000 elephants live in the greater Makgadikgadi, Nxai Pan and Boteti region.

We now no longer have to preserve lion spoor under rocks. Guests at our camps don't have to struggle to find animals. The region is thriving and we have a diverse wildlife ecosystem with southern African desert-adapted antelope including oryx, springbok, plains zebra, wildebeest, steenbok and kudu, as well as their predators – lions, cheetahs, jackals, brown hyenas and more.

From zero lions not too long ago, I estimate there to be more than 100 in our concession and across the unfenced boundary in the eastern sector of the Makgadikgadi Pans National Park. There are days when guests see more than 20 different lions. Some lions will follow the zebra migrations out of the area when the grasses dry during winter, however, and the lion population may decrease to around 50 resident individuals.

Makgadikgadi Pans National Park, Botswana.
© Jose Cortes III

Makgadikgadi Pans National Park, Botswana.
© Super Sande

Zebra migration, Makgadikgadi Pans National Park, Botswana. © Colin Bell

Central Kalahari Game Reserve, Botswana. © Dana Allen

Death at a waterhole

Desmond Clack

Early one morning in Chobe National Park, we encountered a lioness drinking at a watering hole while five others feasted on a young buffalo nearby. As we watched, an elephant cow appeared on the hill, accompanied by a rare sighting – twin calves.

She approached the water with her twins in tow and only when she was almost there did she notice the predator. She trumpeted loudly and in doing so alerted the other lions, who, sensing an opportunity, abandoned their meal and began to approach.

Realising what she'd done and sensing her peril, the cow gathered her twins and started to retreat, but the lions were already closing in.

In a desperate attempt to protect her young, the mother elephant charged at the nearest lion, leaving the twins momentarily exposed. One of the calves stumbled, and the pride pounced, killing it instantly. The mother, frantic with grief, rushed back to her remaining calf, where another large male was already closing in. With fierce determination, she fended off the predator while keeping her surviving calf close, slowly moving them both to safety.

Though she had lost one baby, her selfless courage ensured that the other was saved. In the unforgiving wilds of Chobe, we had seen the power of a mother's resolve.

In Chobe National Park, Botswana, an elephant cow tries to protect her calves from approaching lions.

BOTSWANA

35

Eye-cows and the value of lateral thinking

A simple idea that could deter prowling lions from attacking livestock.

Dr Neil R Jordan

Human–wildlife conflicts are a major source of lion mortality in many parts of their range leading to the decline of large carnivore populations generally. Killing lions in response to attacks on livestock, or even in anticipation of them, is a major challenge to conservation. So it's crucial to reduce livestock losses or compensate farmers in order to retain lions in our shared landscapes. This requires significant effort, resources and ideas.

Non-government organisations (NGOs), local governments and farmers themselves all have a role to play in this, but how do you deter these formidable hunters – the queens of the savanna – from their natural hunting behaviours?

At Botswana Predator Conservation, we tend to focus on separating livestock from lions, commonly through fencing or lethal control, but this is not always possible or desirable. Protected areas are often not large enough to sustain viable lion populations and lions living in surrounding non-protected areas are absolutely key to the viability of the 'protected' populations.

Also, in some parts of Africa, cattle are generally left to wander unattended during the day, before returning home (usually) to a basic livestock enclosure where they spend the night. To some extent, these enclosures can be reinforced with high walls, the addition of barbed wire or thorn branches and further fortified by adding flashing 'lion lights' (a wonderful and effective approach originally conceived and used by Kenyan schoolboy Richard Turere).

But what if the threat of lions is highest during the day, when the cattle graze unattended far from the safety of the enclosure? This is the situation I found at cattle posts on the edge of Botswana's Okavango Delta wildlife area, a global stronghold for lions that abuts local small-scale farming communities. We needed a different approach.

At a particularly remote cattle post in the corner of the veterinary cordon fence, where linear wires separate wildlife and protected areas in the north from livestock-dominated areas to the south, lions were eating livestock at an alarming rate. Originally built to separate domestic cattle from Cape buffalo

Two local BaYei and their dogs return to the relative safety of their village after grazing their cattle on the fertile but lion-infested plains just inland from the Okavango Delta's northern fringes.

Above and right There have been many attempts to find ways to reduce livestock losses due to lion predation around the fringes of the Okavango Delta, Botswana. One of the experiments to find a non-lethal solution was to paint eyes onto the rump of cattle, in an attempt to dupe lions into thinking that cattle were keeping an eye on them. This experiment led to spectacular short-term results.

carrying foot-and-mouth disease, the fence was not intended as a hard barrier for all wildlife. Lions and other carnivores frequently pass over, under and through it to hunt within the livestock areas. As a result of these 'incursions', Ronnie, the owner of this remote cattle post, had lost seven head of cattle in just a couple of months – a rate of almost one a week. Remarkably, in spite of his being a retired professional hunter, he was looking for a non-lethal solution to this problem; he wanted to live with lions. Around the same time, I was mulling over an idea but was not sure if it made sense.

While these lions were hunting cattle to the south, I was sitting in my open-top Land Rover in the shade of an acacia tree watching another lion hunt an impala. The amber eyes of the lioness were laser focused on her prize and, with her whole being pressed almost flat to the ground, she squirmed closer in controlled bursts as the impala shifted from vigilance to browsing.

With painstaking patience she inched closer to her quarry, but something – perhaps a slight twitch – gave her away. The prey registered immediate recognition, communicated through an alarm snort. With that, the lioness relaxed. The mass of ripped muscle and tensed tendons slackened and melted into rest. The game was up.

Lions are predominantly ambush predators, so merely being seen by the target prey was enough to scupper the hunt. This was the seed of the idea, considered a stupid and unfundable idea by some of the main funders of lion conservation, but one I decided to try anyway and which Ronnie was happy to support.

In collaboration with colleagues at Botswana Predator Conservation, University of New South Wales PhD student Cam Radford, Ronnie and later several local farmers, we tested to see if we could employ this natural trick. We decided to paint large eyes onto the rumps of cattle in order to dupe lions

CLAWS is a Botswana-based NGO that collaborates with local communities in the northern part of the country to fund and implement practical solutions that reduce the risk of free-ranging lions attacking and killing livestock. One of their key initiatives is communal herding and night-time kraaling, which helps protect livestock during the hours when hungry lions are most active.

into thinking the cattle were keeping an eye on them. Would this make them give up the hunt?

Ultimately, the lack of resources may have been one of the project's strengths. Taking a thrifty approach, we sourced our inexpensive and locally available materials at the hardware store and other nearby sources.

Armed with paint pots, brushes, foam and wooden boards, we set about making eye-shaped stencils and pressing these to the rumps of more than 600 cattle from 14 herds over the next couple of years. In each herd, one-third of the cattle were left unpainted, one-third got a painted cross as a control and the lucky third were adorned with large eye patterns either side of their tail, which then looked like a long nose or trunk. It seemed straight out of the story of the elephants in the *Babar* books, which painted each other in this way to deter the rhinos.

The outcome was spectacular. Not a single cow or bull with painted eyes was attacked during the study. That such a simple and cost-effective technique could be so successful went beyond our expectations. Here we had a low-cost tool that actually worked.

Following initial testing of the idea in Botswana, several organisations took to using the eye-cow method in different contexts. I am aware of around seven different projects using or having used this procedure in six countries (Argentina, Botswana, India, Kenya, Tanzania, Zambia) to deter at least three species of large carnivore, including lion, leopard and jaguar. The technique has also been applied to a range of livestock other than cattle, including goats, sheep, donkeys and even water buffalo.

Problem solved? Not quite. While the tactic itself seems highly effective across this diverse range of contexts, there are some common challenges around paint longevity and independent user uptake.

Painted eyes tend to last about three weeks and possibly less in very thorny and wet conditions. Work on evaluating paints, patches or other marks would be invaluable here. Surprisingly, we have also had limited direct uptake by farmers, despite efforts to keep cost to a minimum, even providing kits and producing and distributing educational materials in English and Setswana. This also seems to be an issue in other projects, and clearly needs to

> ... to really make progress in conservation, we need an openness to new ideas and a willingness to try different things.

be understood and addressed if we want to see the uptake of this useful tool.

There are, of course, further possible limitations to the technique. First, we have currently always had unmarked cattle in the herd. If you close one burger restaurant but there's another next door, people are still going to eat burgers. In the context of lions and cattle, we don't know whether the lions would take eye-cows in the absence of unmarked cattle but, until further research has been undertaken, it may be most pragmatic to leave a few 'sacrificial lambs' and paint the highest-value individuals.

It is also important to consider habituation. Lions may soon get used to and eventually ignore the deterrent. This is a fundamental issue for nearly all non-lethal approaches, but since peaks in lion attacks on livestock are often somewhat seasonal, it might make sense to apply the technique during only the highest periods of risk.

I hope that, by building a collaborative network of users, we can answer a few of these outstanding questions and resolve some of the issues. The more we discuss and share conservation challenges, the more likely they are to be resolved.

Solutions come from many sources, whether they are simply eye-catching, effective, or both. But to really make progress in conservation, we need an openness to new ideas and a willingness to try different things.

Bulky collars warn shepherds that a prominent lion is heading their way, which has led to a drastic reduction in deaths of livestock. © Colin Bell

To collar or not to collar

Colin Bell

CLAWS is a community and predator-focused NGO that works along the northern fringes of the Okavango Delta. One of its areas of operation is NG12, a beautiful 100,000ha concession that lies to the north of Botswana's Buffalo Fence.

For decades it was designated cattle country, with the veterinary fence creating the solid barrier between wildlife to the south and remote rural villages and their livestock to the north. The fence was erected in the 1970s to separate wildlife from cattle so Botswana's beef could be exported to Europe free of endemic wildlife diseases.

Fast forward to today, where the fence is in total disrepair and has been so for over a decade. This has resulted in wildlife migrating northwards to repopulate the open savannas of NG12, which has fully rewilded and enjoys excellent wildlife concentrations coupled with great scenic beauty.

It's not surprising that safari tourism now dominates the area through a number of recently opened photo safari lodges, providing much-needed jobs and revenues for the neighbouring communities. These help to ensure little incentive to repair the fence and return the area to cattle.

Livestock, in turn have, moved well north and are now found more in NG11 in and around the five villages that lie outside of the Okavango Delta. With large concentrations of prey, the predators too have moved into NG12 in large numbers. Without a hard boundary fence, cattle are easy prey. This has led to retaliatory killings and mass poisonings of lions.

CLAWS stepped in with innovative mobile cattle kraals (bomas) that provide a safer night-time refuge for livestock. They have also collared some of the prominent lions with satellite-tracking devices and an early-warning system that advises livestock shepherds when lions are heading their way. The result is that cattle deaths from lions have plummeted, as have revenge killings; and poisonings have ceased.

The flip side is that NG12 is now a prime wildlife viewing destination attracting visitors from around the world who wish to view lions in their natural state. But tracking collars requiring bulky, highly visible batteries counter the feeling of being in pristine nature.

There is a technology solution. Self-charging pendulum mechanisms have been powering wrist watches for centuries, but surprisingly this hasn't been applied to wildlife collars. This technology could generate energy from the loping movement of a walking lion (or elephant) and eliminate the need for large batteries.

That device could potentially be not much bigger than a wristwatch or even an electronic luggage-tracing tag that could be hidden behind the ear of a lion (or elephant, leopard or cheetah). This would do away with bulky, highly visible collars that must at times be uncomfortable and annoying for the animal. Until that happens, collared lions will continue to be an eyesore for keen photographers, but safety devices for lions themselves.

A Ntsevu sub-adult watching a herd of buffalo at Londolozi Game Reserve, South Africa. © Robbie Ball

The lions of South Africa

PORTFOLIO

Sand River, Londolozi Game Reserve, South Africa. © Kyle Gordan

Londolozi Game Reserve, South Africa.
© James Tyrrell

The Kambula pride, MalaMala Game Reserve, South Africa. © Russell MacLaughlan

MalaMala Game Reserve, South Africa.
© Pieter van Wyk

Kariega Game Reserve, South Africa.
© Brendon Jennings

MalaMala Game Reserve, South Africa. © Jenny Brown

Lapalala Wilderness, South Africa. © Liam Todd/Noka

SOUTH AFRICA

36

Bringing back the lions of Lapalala

Building a base for a population that became locally extinct has been a slow but highly successful project in the magical Waterberg Biosphere.

Peter Anderson

Lapalala Wilderness is a 43,300ha Big Five conservation and wildlife reserve close to the Tropic of Capricorn within South Africa's Limpopo province. The reserve is about a two-and-a-half hours' drive north of Johannesburg and located in the heart of UNESCO's Waterberg Biosphere. Lapalala is a renowned, well-managed wildlife reserve and conservation legacy.

Its two main river systems – the Palala and the Blocklands – rise in the Waterberg ranges nearby and flow north through Lapalala for 63km and from there into what Rudyard Kipling described as the 'great, grey-green, greasy Limpopo' that forms the border with Botswana, some 80km to the north.

The region and the reserve have a fascinating human history, which has seen the landscape shared by humans and animals, including carnivores such as lions, for thousands of years. The large sandstone formations retain groundwater year round; this, together with cliff overhangs offering natural shelters, made it a suitable environment for primitive humans. The discovery of the fossil remains of *Australopithecus africanus* at Makapansgat, only 40km away, attests to the fact that early hominids walked these plains as far back as three million years ago. The San people entered the Waterberg around 2,000 years ago and left behind a number of well-known rock-art sites within Lapalala and its surrounds depicting rhinoceros, numerous antelope and feline forms.

The first European settlers were present in the area in the 1850s and, by the end of the Anglo–Boer War in 1902, all of the lion, elephant, rhinoceros and the remaining big game had been shot out. Eugène Marais, one of the country's greatest naturalists, moved to the Waterberg in 1906 and discovered that the once great wildlife herds were merely memories in the minds of old men. He wrote: 'The Waterberg was the ideal theatre of manly venture, of great endeavours and the possibility of princely wealth. Ivory was then what gold and diamonds became afterwards, and stories were told of bold lucky hunters killing 20 tuskers in one morning, the value of a principality of land in a few hours.'

Richard Wadley, a Waterberg resident, geologist and author of *Waterberg Echoes* wrote: 'For at least the last century, there seem to have been no lion

Lapalala Wildnerness, South Africa. © Dana Allen

resident on the Waterberg plateau; however, there have been occasional accounts of vagrant lion passing through the area. For example, about eight years ago, two lions were seen a few times to the west of Lapalala. Inevitably, this being the backwoods of South Africa, they were tracked and pursued. It's rumoured that one was shot and that the other escaped, perhaps towards the Limpopo River. Certainly, the area north of Lapalala, past the small town of Marken, was still quite well populated with wildlife even into the 1930s when it was a popular winter hunting destination. So, it's possible that lion were still resident, or frequent visitors there.'

The rewilding of Lapalala

Lapalala's rewilding started when Eric Rundgren, from a seasoned hunting family in Kenya, bought the core area of 4,450ha in the 1960s. In 1981, this nucleus piece of land was bought by Dale 'Rapula' Parker, a conservation-minded businessman from the Western Cape, along with renowned wildlife artist and conservationist Clive Walker. They set out with a passion for protecting biodiversity and to create a vibrant, impactful environmental education centre for children. The Lapalala Wilderness School was established shortly thereafter – and continues its work to this day.

It was recognised early on that the next step for Lapalala in order to enhance and protect the pristine wildness of this special place was to increase the size of the reserve to include the river and catchment systems of the Palala and Blocklands rivers. Over time, large tracts of ecologically valuable land and irreplaceable biomes were added, step by step, to reach the 43,300ha of today. The reserve became a sanctuary for rhinos in the early 1980s, when the first critically endangered black rhinos were relocated there. This was a first for a private-sector conservation initiative, setting a new course for what is a UNESCO declared biosphere.

All the species that occurred here historically have been returned, including threatened species like roan and sable antelope and disease-free buffalo that have been bred up and released into the reserve. All of the large and small carnivores that existed historically, including jackal, spotted hyena, brown hyena, leopard, cheetah and lion are now present in Lapalala, which is a fully functioning Big Five game reserve and an important wildlife sanctuary in South Africa.

Lion reintroductions

Lapalala's lion reintroduction programme started in December 2018 with the reintroduction of a four-year-old male and three year-old lionesses. Another four females and one male from Khamab Kalahari were brought onto the reserve in July 2019 to help strengthen the gene pool. The last two male lions came from the Selati Game Reserve near Kruger National Park and were released onto the reserve in November 2022.

Lapalala has had 23 cubs born on the reserve to date. In a bid to control the burgeoning lion population, Lapalala has implemented a range of contraceptive procedures. Dr Annemieke Müller, who heads up the veterinary services at Lapalala, explains the need for interventions: 'Even though Lapalala is over 40,500ha, it has finite boundaries and thus a finite lion-carrying capacity. There is insufficient range for continually establishing new lion prides, even within a large game reserve like Lapalala. A key aspect of managing lions in finite reserves involves switching male coalitions and females between other reserves on an irregular basis to minimise inbreeding with relatives and to maintain typical territorial tenure lengths.

'With a gestation period of 90 days, combined with a high number of cubs per litter, there's a need to practise contraception to keep the lion population within the Lapalala reserve in balance. The process involves the removal of one uterus horn to decrease the space for implantation of fertilised eggs. Lapalala's lionesses will still have cubs, but the litter sizes are smaller with one or two cubs as opposed to between four and five in a litter. One of Lapalala's females has also received a contraceptive deslorelin implant – a reversible treatment that prevents pregnancy much like birth control pills do. Furthermore, to help keep its lion population in balance and to help prevent in-breeding, five lions were darted and moved to the Meletse Game Reserve in August 2022.'

Lapalala has had 48 veterinary interventions, ranging from surgery procedures such as hemi-

Lapalala Wildnerness, South Africa. © Dana Allen

hysterectomies to radio tracking collars being fitted, relocations implemented and tuberculin inoculations. The collars that have been fitted to some of the lions and have been an important tool to aid the research projects that were initiated through the University of Pretoria and the Nelson Mandela University. It has been key for the reserve's managers to understand prey species since reintroduction, as well as the age classes of the prey, the areas in which the lions are hunting, the dynamics of their movements around the reserve and competition and contact with other lions as well as other predators. The planned extensions of the reserve into more ecologically valuable habitat will ensure that the requirements to manage the lion populations may be reduced.

Lions as an indicator species

Lions are the apex predator in any African wildlife national park or game reserve and Lapalala is no exception: they're an important indicator species as to the health of the reserve. A healthy lion population can be associated with the generation of sustainable landscapes, which provide a wide range of benefits to other species in the ecosystem, especially antelope and other ungulates. Lions remove the sick and weak animals and can prevent the transmission of diseases in herbivore populations. They play a crucial role in maintaining a healthy balance of wildlife numbers in any reserve. An absence of lions can most certainly

Lapalala Wilderness, South Africa. © Jack Cooper

lead to a dysfunctional landscape, where smaller carnivores are not capable of regulating larger herbivore populations.

From an evolutionary viewpoint, the use of fences to create large-scale game reserves for conservation – as has happened around southern Africa – has not affected the natural behaviour of lions as they still conform to expectations derived in unfenced reserves. Prey abundance is the key factor in determining the use of space by lions and is similar in both fenced and unfenced reserves.

Lion management

South Africa's Biodiversity Management Plan for African Lion was published in 2015 in terms of the National Environmental Management: Biodiversity Act (NEMBA) by the Minister of Environmental Affairs. Lions in South Africa will provide key opportunities for biodiversity conservation, economic development and social benefits through the existence of stable, viable and ecologically functional populations consisting of wild and managed wild lions.

In terms of South Africa's lion management plan, Lapalala's lion population is classified as 'managed' wild lions as they occur within a closed (i.e. fenced) reserve and are managed to limit unsustainable population growth and maintain genetic diversity. This means Lapalala's management team have to actively manage lion demographics.

Lions eat a wide range of mammals, tending to favour medium to large ungulates with an average body mass of 190–550kg and are known to scavenge whenever possible. Lions respond to behavioural and physiological changes in prey in terms of what species, sex and age of prey they select. Typically, in an area like Lapalala, some five ungulate species make up the bulk of the lions' prey. This pattern occurs in both large open systems and in smaller fenced reserves.

Pride ranges respond to changes in food abundance on an annual rather than a seasonal timescale. Thus, female home-range size is driven by the size of the pride and by prey abundance. Lions hunt most successfully in areas of thick cover and long grass and especially when there's no moon. Additionally, when lions are within two kilometres of a waterhole, they move more slowly, cover shorter distances at night, and follow a more tortuous path than when they are further from a waterhole.

The way forward

Rapula Parker, the man of rain, passed on the management of Lapalala – an enormous responsibility – to his son Duncan, who was joined in 2013 by Gianni Ravazzotti. They have set Lapalala on course for the next stage to being an exceptional role-model conservation legacy and to ensure Lapalala is a world leader in wildlife protection and conservation.

To help make the reserve sustainable, a small number of high-end, low-volume and low-impact lodges have been opened to attract and host discerning wildlife enthusiasts. The Lapalala Wilderness School has expanded its reach into the adjacent communities as well as schools further afield and has hosted over 150,000 schoolchildren since inception. Lapalala has been the birthplace of the Endangered Wildlife Trust (EWT), the Rhino and Elephant Foundation (REF), the Field Guides Association of Southern Africa (FGASA) and many research and conservation initiatives.

It is clear that lions and primitive humans have shared this landscape for tens of thousands of years with seemingly minimal impact on lion populations. In the last 200 years, however, lions have been annihilated by humans, with local extinction occurring around 120 years ago. Despite this, wild, free-roaming lions have continued to find ways into this large area, aided by rewilding in the last 60 years. Lions are now permanently entrenched once again.

With lions now found on less than 8% of their original range across Africa, Lapalala's rewilding model is there to inspire and encourage protected areas of all sizes to look to adding new land or joining with other protected areas and increasing the range for these tawny cats. The importance of lions and the role they play in balancing ecosystems cannot be overstated.

SOUTH AFRICA

37

The rewilding of Kwandwe

Kwandwe is isiXhosa for place of the blue crane, the national bird of South Africa.

Ryan Hillier

The founding of Kwandwe Private Game Reserve was serendipitous. Carl DeSantis and Erika Stewart were vacationing in the Linyanti region on the northern border of Botswana in the late 1990s. They had elected to stay at King's Pool where they met and befriended the camp managers Angus and Tracy Sholto-Douglas, a South African couple who had grown up in the Eastern Cape and wished to return there. Chatting around the campfire, they discovered a shared interest in creating a new premier Big Five game reserve.

The relationship that started on the banks of the Linyanti River meandered to the Great Fish River Valley in the Eastern Cape, resulting in the birth of Kwandwe Private Game Reserve. With the purchase and amalgamation of many old farms, the 75,000ha reserve is widely recognised as one of South Africa's premier private game reserves and a shining example of a highly successful rewilding project that benefits both biodiversity and people. The reserve now employs over 250 people, with a strong focus on rural community development through their Ubunye Foundation.

In 2012, the Chouest family from the USA purchased Kwandwe and have driven the game reserve to even greater heights, with well-supported infrastructure and financial sustainability. Kwandwe today is even more focused on endangered species protection, most notably black rhinos, but also the large predator populations that now prowl Kwandwe after 150 years of local extinction.

Kwandwe's founding lion population originated from the Madikwe and Pilanesberg game reserves in 2001. The initial four lions, two males and two females, were successfully introduced and are the ancestors of the thriving prides on Kwandwe today. A total of 44 litters of lions have since been born here. The periodic incorporation of new genetics in the form of introduced males, as well as the ecotourism-friendly nature of the lions, has made the genetically diverse population on Kwandwe highly sought after. A highlight is the translocation of lions to areas of local extinction further north in Africa, such as supplying a founding population of lions to Liwonde National Park in Malawi – a good example of how South Africa's lion population can be used to repopulate conservation areas further afield. Kwandwe has translocated lions to 25 other reserves in Africa, playing an important role in increasing the distribution and genetic diversity of wild lions.

Kwandwe's accurate record keeping over the years has become a model for conservation-management strategies for lions, and the data collected has created multiple research opportunities. To date, Kwandwe has contributed to over 20 peer-reviewed publications on lions and other large carnivores.

Kwandwe Private Game Reserve, South Africa. © Ryan Hillier

Kwandwe Private Game Reserve, South Africa. © Ryan Hillier

The Great Fish River meandering through Kwandwe Private Game Reserve, South Africa. © Ryan Hillier

The rewilding of a landscape as diverse and important as Kwandwe brings with it a responsibility for its guardians in the future. A thriving lion population is not only a beacon of hope for lions as a species, but for all wildlife that walks this ancient landscape. There will be no lions without the balance and success of the whole ecosystem.

SOUTH AFRICA

38

Rewilding farmland for lions

Kariega Game Reserve shows what can be achieved in the conservation of lions when people come together to find innovative solutions.

Lindy Sutherland

The Bushman's River Valley in South Africa's Eastern Cape province is today a place of great beauty and biodiversity, but it was not always so. Over 200 years ago it was settled and converted from natural bush to croplands. Around 35 years ago that began to change when the first farm was bought and rewilded.

At that time all lions, and indeed almost all wildlife, had been shot out. Over time, another 22 farms were bought and woven together into what is now the Kariega Game Reserve, a treasure of great ecological value. And the lions are back.

The reserve covers more than 10,000ha of fully protected wilderness, a shining example of what can be achieved when people work together to create protected habitats for wild animals.

Many species have been re-introduced, including lions, elephants, buffalos, cheetahs, hippos and white and black rhinos. As the conservation area matured, many other animals also reappeared to fill their niches, such as caracul, aardvark and even oxpeckers, which had become locally extinct throughout the province.

Since reintroducing lions in 2004, Kariega has contributed significantly to lion conservation. As members of the Lion Management Forum, it collaborates with other protected areas in South Africa to share experience, best practices and genetics, while participating in research projects that determine how best to mimic natural processes and ensure sustainable populations.

The long-term vision for the region is to create over 200,000ha of unfenced wilderness landscape. To make that happen, more and more Eastern Cape farmland needs to rewilded and linkages made to game reserves with viable corridors. This would stitch together the whole area into a wild fabric where the future of lions and other wildlife is secure.

> The reserve covers more than 10,000 hectares of fully protected wilderness, a shining example of what can be achieved when people work together to create protected habitats for wild animals.

Bushman's River, Kariega Game Reserve, South Africa.
© Brendon Jennings

Kariega Game Reserve, South Africa. © Brendon Jennings

Kariega Game Reserve, South Africa. © Brendon Jennings

Kariega Game Reserve, South Africa.
© Brendon Jennings

© Dr Andrew Baxter

SOUTH AFRICA

39

Lions return to Babanango Game Reserve

For the first time in nearly 150 years, lions are once again roaming the spectacular rolling hills and valleys of Babanango Game Reserve in KwaZulu-Natal.

Chris Galliers, Conservation Outcomes

The reintroduction of lions into this landscape – after a very long absence – required much planning and careful execution. This started in 2018 with its transition from an environment that had been overutilised for agriculture to one that, in just five years, would be home to KwaZulu-Natal's newest Big Five reserve. What's more, most of the 20,000ha is owned by three of the local communities.

This new game reserve realises the local Emcakwini, Esibongweni and KwaNgono communities' decades-long vision of creating a wilderness area that would attract tourism and stimulate the local economy. Their vision became a reality in late 2017 when Hellmuth and Barbara Weisser, philanthropic German investors who share a deep passion for wildlife and conservation, started to purchase land and, together with the three local community trusts, created a project that was to become the Babanango Game Reserve.

To get to the point where a lion population can be introduced requires much work, support and funding. Firstly, all 20,000ha of the reserve had to be fenced to the legal specifications for predators and elephants, which amounted to 80km of electrified game fence,. We then needed to introduce a prey base that would have been endemic to the area and, over a number of years, a total of nearly 3,000 animals were introduced made up of 21 different mammal species.

A Predator Management Plan was required by the provincial agency to detail how the lions would be introduced, and at what ratio, taking into account that there is a carrying capacity for predators in a closed system. A specific lion boma was constructed in preparation for the new arrivals which were to be a coalition of two males and then four females, of which two would be from one pride (and reserve) and the other two from another property so as to ensure a genetically diverse founding population.

Lions play a crucial role in the African ecosystem as apex predators, fulfilling an important ecological

White Umfolozi River, Babanango Game Reserve, South Africa. © Colin Bell

role on the reserve. Grazers, such as large herds of ungulates like eland, zebra, wildebeest and buffalo, are growing in abundance at Babanango Game Reserve and predation by lions will help manage their populations by removing old, sick, or injured animals, thus promoting healthy populations. Additionally, the presence of lions benefits other scavenging species, such as vultures and brown hyenas, by providing access to carcasses.

In February 2023, the lion permits from the conservation authority were issued, enabling Babanango Game Reserve to proceed with introducing the lions. Two young males born on Nambiti Private Game Reserve in KwaZulu-Natal were caught in June 2020.

Initial efforts to capture them failed as they were in dense cover and could not be darted. They were nervous as a result of being chased by their fathers. After many hours on the hunt, the team eventually located them using very high frequency (VHF) telemetry, but to catch them was not going to be easy as they were in a block of land with long grass and poor accessibility. Under the guidance of veterinarian Dr Ryan van Deventer (from Wildlife Solutions Africa) and me, it was decided that they would lure the young lions with a culled wildebeest laced with sedatives. After about two hours, the team returned to the carcass and were happy to find the lions sedated, having feasted well. They were then loaded and transported for their release into the lion boma at Babanango Game Reserve the next morning (18 March 2023).

'Our reserve is situated in the heart of Zululand, near where seven Zulu kings are buried in an area called "The Valley of The Kings". The introduction of "The King of Beasts" into Babanango Game Reserve was barely imaginable a mere three years ago but now that it has come to fruition, we are ecstatic and one step closer to fulfilling this incredible dream of Big Five status,' said Musa Mbatha, Babanango Game Reserve manager at the time.

During the pre-release phase, the two young males, which were in excellent condition and differentiated by the colour of their manes, one dark and the other light, were held in the Babanango lion boma for monitoring before being released onto the reserve. Before the release, both lions were fitted with satellite tracking collars to assist in post-release monitoring, with their movements being tracked daily to see how they use the landscape over time and also to mitigate risk in the event that they might break out.

Introducing lions as one of the final species in the rewilding process of Babanango Game Reserve is no small feat. Essentially, one is bringing a large, dangerous predator into a landscape where, for centuries, humans tried to exterminate them because of the threat they posed to lives and livelihoods. But it was ultimately successful.

Following the successful release of the two males, two lionesses were introduced to the Babanango Game Reserve lion boma in July 2023. The two young cats, from the same pride and born in 2021, were caught at Kwandwe Private Game Reserve in the Eastern Cape. They were darted by Dr Van Deventer and loaded into a specialised trailer for transport to the reserve. Once safely in the boma, both cats seemed really relaxed and comfortable and, after being collared, were released into the park in September 2023 to make way for the next intake.

The next two lionesses were sourced from Mount Camdeboo Private Game Reserve in the Karoo, whose lion population at the time was over reserve capacity. The lionesses (one of 18 months and the other four years) were from the same pride and shared the same father. Steve McCurrach from the Bateleurs – volunteers flying for the environment – and Dr Van Deventer drew up plans to fly the

> To hear the sound of a lion's roar reverberating through the valleys of Babanango Game Reserve, the heart of Zululand, will be the sound of another conservation success in this incredible project.

lionesses from Mount Camdeboo to Babanango Game Reserve. Lincoln, a volunteer with the Bateleurs, offered his plane for the mission.

On 13 September 2023, the plane left Virginia Airport in Durban en route to the airstrip at Asanta Sana, which neighbours Mount Camdeboo. However, due to high winds, a decision was made to land at Graaff-Reinet airstrip, a better landing strip. The Mount Camdeboo team met Dr Van Deventer and Lincoln and took them through to the reserve. The two lionesses were darted and weighed, with their combined weight amounting to 280kg. However, the plane had a weight limit of 235kg. This, together with high winds, meant that it was necessary to take extra precautions, and the cats would have to be transported one at a time. The younger lioness was loaded and flown to Ulundi airport where the Babanango Game Reserve team was waiting to transport her to the boma.

The severe weather associated with a frontal system then prevented the team from transporting the other lioness to Babanango Game Reserve for several days. On 18 September, however, when the weather had improved, the second, larger (150kg) lioness was flown to Ulundi, and all went smoothly. She was now also released into the boma and the two big cats were reunited, thankfully with no aggression. They remained in the boma for about six weeks before being released into the reserve and have been thriving in their new home ever since.

The reintroduction of these iconic predators to this region of Zululand is part of a larger vision for the reserve and its contribution to conservation, and is a significant milestone for conservation in South Africa – lions are listed as Vulnerable by the International Union for Conservation of Nature (IUCN) Red List of Threatened Species. Key threats to lion populations in South Africa include habitat loss, human–wildlife conflict, poaching, disease outbreaks, and the illegal wildlife trade. These threats contribute to the decline of lion populations and require ongoing conservation efforts to ensure their long-term survival.

The stability of lion populations relies on several key factors, including the growth of their population, maintenance of genetic diversity and, most significantly, the preservation of their natural habitat. The rewilding of Babanango Game Reserve is part of management's commitment to conservation and the preservation of the natural habitat. Over the past five years, the reserve has undertaken an extensive game translocation initiative, sensitively reintroducing close on 2,500 large mammal species back into the area, where once such species roamed freely. This successful endeavour has included the reintroduction of endangered black rhinos and rare antelope like oribi and klipspringer. In addition to this, and since August 2023, a herd of 17 elephants now roam these hills for the first time in 150 years.

The reserve incorporates a significant portion of land that has been invested by three community trusts from the region, whose commitment plays a crucial role in long-term conservation success. The income generated through land leases, conservation levies, and other benefits from the reserve is vital for the economic wellbeing of the surrounding communities.

To hear the sound of a lion's roar reverberating through the valleys of Babanango Game Reserve, the heart of Zululand, will be the sound of another conservation success in this incredible project.

Preserving Makuleke

Before 1969, the northern boundary of Kruger National Park stopped at the Luvuvhu River, around 10km south of the Limpopo River, which demarcated the boundary between South Africa and Zimbabwe. The conservation authorities of that time decided that the community lands that make up the Pafuri Triangle between the Luvuvhu and the Limpopo rivers should be incorporated into the Kruger National Park because of its extraordinarily rich biodiversity.

While comprising only about 1% of the Kruger National Park's total area, the Pafuri Triangle area contains plants and animals representing almost 75% of the park's total diversity. In 1969, the community of around 1,500 people living in the Pafuri area were forcibly moved off their lands and resettled outside of the park. Kruger's northern boundary was then moved northwards to the Limpopo River.

After the first democratic elections in 1994, anyone who had been forcibly removed off their lands after 1913 could claim to get their land back. In 1996 the Makuleke people submitted their land claim for the 17,400ha that make up the Pafuri Triangle region. In spite of 'progressive' new policies of the South African National Parks (then the National Parks Board) and top-level political support for the Makuleke claim, a protracted bargaining process ensued between the community and the conservation authorities. In the face of considerable objections within the park, the Makuleke people were formally handed back their land at a signing ceremony in 1998. Many predicted that this would be the end of conservation in South Africa. How wrong they were proved to be!

The community elected not to resettle on their land but to engage with the Kruger Park and the private sector to create a contractual park that they owned and that would be incorporated back into Kruger's boundaries. They elected to partner with the private sector to invest in tourism, thereby creating much-needed jobs and revenues for their community. That decision resulted in the building of several game lodges with the community as the prime beneficiary, a win-win for biodiversity conservation and community upliftment. The Pafuri area is today one of the most magical areas in a world-class park.

Makuleke Contract Park,
Kruger National Park, South Africa.
© Dana Allen

SOUTH AFRICA

40

White lions

Strange feline ghosts are an enigma that could not look more out of place in the earthy shades of an African savanna.

Chad Cocking

It was a sweltering early summer's afternoon in late 2010 in the Timbavati Private Nature Reserve – part of the 35,000km² Great Limpopo Transfrontier Park (GLTP) – and I was watching a couple of spotted hyenas cooling off in a small waterhole. The day before, a few hundred metres from the water, a compact, sleepy pride of lions had lain, their bellies bulging from having gorged on a sizeable male giraffe over several days.

By the following afternoon maggots were more plentiful than the remaining flesh and the six members of the Xakubasa Pride were resting, stomachs engorged. As the sun sank beneath the western horizon and a relative coolness descended over the bush, the lions failed to rouse themselves to drink and we had only the hyenas as entertainment. With the lions around, nothing else was daring to drink. Then the radio crackled: 'The *ngala* (lions in Xitsonga) have woken up.' We sat tight.

The hyenas that had been milling about suddenly took off. The reason for their quick exit became apparent; a young male lion emerged in full view of both the hyenas and us. His slightly younger, somewhat different female cousin soon joined him and they looked daggers in the direction of the retreating hyenas. The young lions then padded their way confidently to the water's edge and began to lap. I raised my camera and clicked … and clicked, forcing my hands not to shake. This was a scene I had been dreaming about since I began wildlife photography: the second lion was pure white.

Born with a recessive gene carried by both her tawny mother and father, she had a mutation of the tyrosinase enzyme, which caused a reduced production of melanin (the dark pigment responsible for colouring eyes, skin and hair). This recessive gene-carrying lioness had a decreased deposition of this pigment along the shafts of her fur, which was almost colourless.

A lion without colour is like a ghost; a snowy enigma that could not look more out of place in the earthen shades of the African savanna. Yet here she was, drinking water next to a conventionally coloured lion, which vividly accentuated her physical differences. Her ice-blue eyes locked on me as her pink tongue lapped from slightly pinker lips, but brown patches on her ears (as well as those blue eyes) showed she was not an albino.

Timbavati Private Nature Reserve, South Africa. © Chad Cocking

Timbavati Private Nature Reserve, South Africa. © Chad Cocking

It was a sight never to be forgotten and with only two known white lions roaming the Greater Kruger system at that stage, I knew I had captured a unique sight, captured by no-one before.

The return

In 2010, two white lions were spotted roaming Timbavati, the first seen in almost two decades. The fact that no further sightings of white lions were made during that period resulted in their being classed as 'technically extinct' in the wild, which ignited a drive by some to one day supplement Timbavati's 'extinct' gene pool. Yet, the gene was there; this absence was nothing new – white lions have been absent from our wild systems for far longer periods than they have been present in them over the past century. White lions had found their way into the folklore and oral histories of African elders in the Timbavati region, the first documented sighting having been made by Joyce Little in the Timbavati in 1938.

It was not until 1975 that Chris McBride, a Timbavati member, took the first photographs of white lion cubs in one of the prides he was researching in the area. He had witnessed their conception – again, between two tawny coloured lions – and had been waiting for his first glimpse of the cubs. One day his sister came rushing into their camp, telling Chris he needed to come see something. After arriving at the pride, it didn't take long for two snow-white lion cubs to pop their heads out from the tawny masses. Their golden eyes – as opposed to the blue ones seen in the white lion I photographed – were all Chris needed to see to know that they weren't albinos; we now know the difference in eye colour is a result of differing expressions of the recessive gene that these lions possess.

Seeing how easily they stood out from their surroundings, the question of how well they were going to survive in the wild with such a disadvantageous coat coloration shot to the forefront of Chris' mind. How would they evade detection of predators and enemies as young cubs trying to hide when their mother was off hunting? If tawny-coloured cubs fell victim to these threats on a regular basis, what chance would a pure white lion cub have of staying out of harm's way? If by some miracle they made it through those critical months at the start of their lives, how would they fare as they got older? Surely it would not be possible to be a successful hunter when your coat would give away your presence even on a moonless night.

For these reasons, after a couple of years of observation and worry about their chances, and not knowing if any more of these highly unusual animals would be born, it was decided they would be captured and moved to the Pretoria Zoo for their own safe keeping.

Few would argue that zoos – or any form of captivity – should be places for lions to live out their days. However, in the following decades, the recessive gene that led to the unique coats of the two white Timbavati lions spread across the world. Zoos, circuses and safari parks became home to more and more selectively bred white lions. While there were no white lions alive in the wild at the time, by the late 1990s, it was estimated that around 300 white lions resided in captivity. The worst breeding programmes involved breeding white lions for 'canned' lion hunts, often the cubs were hand-raised by humans, only to be released later into small enclosures with trophy hunters paying a handsome fee to kill them.

White cubs

Fortunately, in the late 1970s, three more white lion cubs were born in Timbavati. One – a female called Phuma – was born in 1976 to the Machaton Pride, which Chris was still studying. In 1981, another two white cubs were spotted in a litter born to an unrelated pride in the northern Timbavati. Of this litter of one male and one female white lion, only the female survived. Unoriginally called Whitey, she proved herself to be the first white lion to show that, despite her seemingly obvious handicap without camouflage, it was possible for a white lion not only to survive but also to thrive in the wild. Whitey's Pride had another white cub (Ntombi) in 1991, but when Whitey disappeared at around 12 years of age and Ntombi was killed by nomadic male lions in 1993, it brought an end to the presence of the white lions in the Timbavati.

As the years passed and no further white lions were born, it was presumed that the gene was

gone. But it hadn't – a recessive gene can pass through tawny populations for several generations before resurfacing. Lionesses typically stay within their natal pride, keeping the genes well concentrated, but young males disperse from their natal ranges, pushed out by the dominant males to become free-roaming, nomadic individuals until they are older, bigger and strong enough to take over a pride of their own. This reduces the risk of inbreeding: a big plus for the genetic diversity of the lions, but a hinderance to the chances of recessive white lion genes pairing up between two gene-carrying lions.

The chances of these recessive genes finding one another in a massive open system and producing a white cub are low. Young Timbavati male lions carrying the white lion gene will disperse into other regions of the Greater Kruger system and male lions without the gene move into the Timbavati where they take over prides. But, given enough time, two white-gene-carrying lions will find one another and mate, to produce a white lion cub. Add to this the statistics that even tawny lion cubs have as little as a 10–20% chance of surviving to maturity, the fact that any white lions survive is remarkable.

Yet, despite the chances, no fewer than seven wild white lions have survived to adulthood in Greater Kruger, including some born 60km away from Timbavati. Some limited studies show that 19% of sampled lions in Timbavati carry the white-lion gene, proving the gene is alive and well and making interventions to protect and proliferate a recessive gene unnecessary.

It's about conservation

While the white lions of the Timbavati are not a separate species and do not need any special status, there is something to be said about ensuring that this recessive gene – which has not appeared anywhere else in Africa – should be conserved. The way to do this is not by conserving white lions per se, but rather by conserving all lions in the region. Like almost all global conservation areas, the Timbavati and the Greater Kruger regions face the unrelenting pressures of human population growth on their borders. This book has already shown that habitat loss (and concurrent loss of prey species) and habitat fragmentation are two of the biggest dangers facing wild lions and other species – and projected demographics show that people will be staggeringly outnumbering wildlife in the years to come.

As part of the GLTP, Timbavati is in a unique conservation space; while many reserves and conservation areas across the world are shrinking, Timbavati and the GLTP are growing. Wild animals need space, and larger conservation areas provide for greater stability of populations and ecosystems, especially in the face of climate change. The bigger the system, the more resilient it is in the face of these unprecedented changes. However, space alone – and what happens within the confines of these protected areas – is not enough.

A vital piece of the puzzle is what happens beyond the borders of these protected areas, most tellingly with communities that live on their doorstep. They need to see, and feel, the benefit of conserving these natural systems to which they often do not have access. They must understand why it is important to protect the species that could potentially pose a threat to them and their livelihoods. Many are extremely poor with more pressing issues than protecting wildlife. Sustainable job creation connected with ecotourism needs to be felt within communities surrounding parks, and tourism operators must ensure there are meaningful local supply chains that carry positive socio-economic benefit.

Despite all the positive gains the ecotourism industry has made in this regard over the past decade (over 10,000 people now find direct employment through tourism operations adjacent to the western boundary of the Greater Kruger), growing human populations will continue to put pressure on these natural systems (there are presently over 2.7 million people on Kruger's western boundary). What will convince these communities to protect wild spaces in order to secure their own economic future and for the intrinsic value of wild space in a world where space is at a premium? In this there is value in the local belief that white lions are totemic, and their presence is a sign of peace and prosperity to all living in these tribal lands, which gives us yet another reason to conserve them.

Sanbona Wildlife Reserve, South Africa. © Ian Malone

SOUTH AFRICA

41

Bones of contention

From 2008 to 2018, South Africa permitted the export of captive-bred lion skeletons. The wellbeing of a lion in captivity is irrelevant when all you want is its bones.

Dr Don Pinnock

Jabula was born on Lion & Safari Park in Hartbeespoort, North West province. When he was just a few weeks old, he was taken from his mother and moved to Chameleon Village where he was monetised for cub petting. The story told to tourists and volunteers was that his mother had died and he needed to be hand-reared until he was ready to be returned to the wild. It was a lie. Eight months later, too big for petting, he returned to his birthplace to be used for tourist walks. There, a special bond formed between Jabula and his keeper, Armand Gerber. They would hug and the lion would roll over for a tummy scratch.

But young lions grow into big, strong predators, which become too dangerous to entertain tourists. At 18 months Gerber discovered that Jabula would be sent away. He began negotiations to buy him. On 22 April 2018, with the purchase still pending, a team from Wag-'n-Bietjie farm near Bloemfontein arrived to collect the lion despite Gerber's protests. The men had presented no permits and no vet was present. Jabula was inexpertly darted and hauled away.

On 23 April, permission was granted for Gerber to buy Jabula. He contacted Wag-'n-Bietjie, but discovered that the lion and his brother Star, who had been removed earlier with 19 other lions, were being held in a small crate with no food or water.

Desperate, Gerber contacted the SPCA in Bloemfontein, asking them to investigate. Inspector Reinet Meyer arrived at Wag-'n-Bietjie on 24 April. Outside a farm shed she noticed a large pile of rotting, fly-covered meat. On entering, she found a supervisor and about eight workers stripping the skin and flesh from the fresh carcasses of 26 lions, including Jabula and Star. Their flesh was being boiled away and they were being reduced to clean skeletons.[1]

That afternoon a truck arrived with 28 additional lions, which were to be killed the next day. They were darted by a vet then shot through an ear with a low-calibre .22 rifle to not damage their skulls. 'Overseas buyers don't want a skull with a bullet in it', he told her, which is why he didn't shoot them directly in the cranium. More than 240 were awaiting the same fate.[2]

Meyer had stumbled on a lion slaughterhouse feeding an Asian demand for lion bones. She filed charges and the police began investigating on an

Trophy hunter, Northern Cape, South Africa. © David Chancellor

allegation of animal cruelty. It was known that South Africa was permitting the export of lion bones, but the scale of the slaughter was unprecedented.

Then the ducking and diving began. According to Albi Modise of the Department of Environmental Affairs, the department could not comment on mass killing because, he said, the welfare of captive-bred lions fell under the mandate of the Department of Agriculture, Forestry and Fisheries (DAFF). When approached, DAFF said the lions weren't their responsibility either, but that of the Free State Department of Economic, Small Business Development, Tourism and Environmental Affairs (DESTEA). According to Dirk Hagen of DESTEA, the primary responsibility of the lions' welfare resided not with him but with the SPCA and the animal owners. Nobody was accepting responsibility for one of the world's most shocking human betrayals of wildlife custodianship in a country that prided itself as a beacon of conservation.[3] Where did this grisly business begin, and why?

Pleasing the hunters

Hunting lions as trophies is expensive, tiring and not without danger. But what serious hunter can consider their expedition to Africa a success without bagging a snarling lion head to hang on their trophy wall? In South Africa, a solution was in the making. Almost uniquely in the world, private citizens are permitted to own and breed wild animals. Given access to land, good fencing and a supply of meat – old horses and battery chickens are considered fine – lions are not difficult to breed.

There were and are many private reserves in the country where they're kept, cared for and are part of private collections for tourist viewing. But, increasingly, game farmers began responding to opportunities presented by trophy hunters with few skills and deep pockets who were willing to bag a lion on a fly-in safari with no risks and a taxidermist to hand. Canned lion hunting was born.

In South Africa, the captive breeding of lions began in earnest in the 1990s and by 1999, there were about 1,000 lions in the industry. Today, there are an estimated 366 facilities and between 8,000 and 12,000 captive lions providing relatively cheap specimens, mostly for foreign hunters. Male lion shoots are said to be sold at between US$25,000 and US$40,000, females half that. There could also be over 1,000 captive tigers.[4]

Lion breeders have learnt to profit from every stage of these lions' lives. Wild female lions give birth only every 18 months to two years, but in captivity cubs can be removed within days of being born, which brings the female into oestrus sooner and allows the females to produce up to four litters every two years. Income and labour is derived from 'voluntourists' (often young foreigners in their gap year) who pay for the chance to hand-raise the cubs while being told that 'their' lions will be released into the wild. Later, adolescent lions are used to accompany awed tourists on 'lion' walks.

The trade in illegal cubs in South Africa is also substantial, with prices ranging from US$600 to US$950 for a cub of one week to four months old, and US$950 to US$1,500 for a tiger cub, depending on the age, sex or subspecies.[5]

Trade data focusing on live lions, trophies and skeletons was explored in a FOUR PAWS report. Between 2008 and 2017 South Africa legally exported more than 16,000 lions and more than 400 tigers, dead or alive, with most sourced from captive populations.[6] Numbers of CITES export permits issued in South Africa for the translocation of lions were consistently higher than permits recorded in importing countries. This is also true for tiger exports.

Because these lions are habituated, have little fear of humans and are often in-bred, captive-bred lions have zero value to conservation or the estimated 3,000 wild lions in South Africa. Their existence is purely commercial. In North West province, where a large percentage of captive hunts take place, the shortest legal release time from captivity until the hunt is four days.

Opposition kicks in

Canned lion hunting first came to public notice when British filmmaker Roger Cook screened 'Making a Killing' in May 1997. Using the cover story of being a hunting outfitter and with hidden cameras, the team infiltrated the Lowveld hunting industry, filming the covert world of 'tame' hunting.

'So-called canned hunting,' he explained, 'involves unfairly preventing the target animal from escaping the hunter, thereby eliminating "fair chase" and guaranteeing the hunter a trophy. The hapless animal is handicapped either by being confined to a small enclosure or because it has lost its fear of humans as a result of hand-rearing. Some are even tranquilised.'[7]

They were told by professional hunter Kerth Boehme: 'He [the client] just has to do a little bit, get the lion to the bait, make a half-hearted attempt to track the damn thing, or just shoot it from a blind [hide],' and that 'even the weakest of clients can still be satisfied and still at least have done it, you know, semi-ethically.'[8]

Then, in 2015, the documentary *Blood Lions* anchored by environmentalist Ian Michler followed, using *The Cook Report*'s technique of booking a hunt to penetrate the industry, uncovering shocking conditions on breeding farms.[9] The film fuelled the growing global antipathy towards lion hunting.

Beyond South Africa's borders, opposition to canned lion hunting was building. Australia and France banned the import of trophies from captive lions and, in late 2016, the US Fish and Wildlife Service – under the Obama administration – followed, saying the industry had not proved it benefited the long-term survival of lions in the wild. Membership of South Africa's main hunting associations (Professional Hunter's Association of South Africa and the National Confederation of Hunters Associations of South Africa) to the European hunting community was revoked. Almost overnight, the industry lost more than half its hunting clients. Captive lion prices began dropping.[10]

One of the world's leading experts on lions, Dr Paul Funston, was quoted as saying: 'It is confounding that a country whose iconic wild lions are such a source of national pride, not to mention tourist revenue, would take such risks as to sustain a marginal captive breeding industry that is condemned globally for its shameful practices.'[11]

Selling bones

It was clear to lion breeders that canned hunting was heading for hard times – because of both local and international opposition. But another avenue was already in place: the sale of lion skeletons to Asian buyers who were passing them off as tiger bones used in traditional medicine and 'tiger bone wine' made from bones soaked in alcohol. While hunted lions required a level of care and good feeding to attract clients, selling their bones necessitated virtually no concern for their wellbeing when alive.

Lion bones had been exported from South Africa since 2008, but in 2017, the Department of Environmental Affairs set a quota of 800 carcasses a year, raised it to 1,500 the following year, then brought it back to 800 after widespread NGO opposition. Given the lack of capacity among South Africa's law enforcement fraternity to check consignments at border posts, ports and airports, cheating on numbers and laundering of wild lions into consignments was inevitable.[12] A time of extreme cruelty to lions had arrived.

Unfortunately, bone sales were aided by a 2016 resolution taken at the 17th CITES Congress of the Parties (CoP17) that allowed South Africa to continue to trade in lion parts from captive lions, effectively creating a split-listing problem.[13] However, on the ground, that separation was almost impossible to police, the resolution was used to justify the setting of bone export quotas into which wild lions and even tiger bones could be leaked.

Dr Kelly Marnewick, formerly Senior Trade Officer for the Endangered Wildlife Trust (EWT) Wildlife in Trade Programme, warned that 'poaching of wild lions for body parts has escalated in recent years and we cannot rule out a link to the market created for lion bones from captive breeding institutions'.[14]

An explosive report in 2018 by the EMS Foundation and Ban Animal Trading, titled 'The Extinction Business: South Africa's "Lion" Bone Trade', found a lack of verification in skeleton numbers as well as a lack of due diligence by CITES authorities, with substantial loopholes in its permissions system.[15] As 91% of the skeletons that went out in 2017 included skulls, 'It can therefore be concluded, contrary to claims from government, that South Africa's lion-bone trade is not simply a byproduct of the canned trophy hunting industry. Big cats are being commercially bred for their bones.'

Writing for the Organised Crime and Corruption Reporting Project (OCCRP), Khadija Sharife quoted

S J Alberts, who had spent more than 40 years working in the lion farming industry: 'I believe that the first mass euthanasia permits were issued around 2016. Some breeders saw the writing on the wall for the captive breeding industry. Suddenly the … value of thousands of animals existed only in their bones.'[16]

In an attempt to increase profitability in the captive breeding and keeping of lions and other big cats, according to FOUR PAWS' 'The Vicious Cycle' report, when it comes to the lion bone trade, animal welfare was not a priority. 'A lack of adequate basic animal welfare conditions, such as sufficient water, food, shelter and medical care, is inevitable and a stark reality in South Africa.'[17]

Bargaining with devils

Lion bones were leaving South Africa, but where were they going and why? According to export records at OR Tambo airport in South Africa, most were consigned to Laos. But according to customs records, none actually entered that country. The OCCRP investigation found they were quietly being rerouted to neighbouring Vietnam.[18] These volumes far exceeded the quota set in South Africa, with shipments either undervalued or stuffed with extra bones.

When its researcher tried to visit the top bone buyers in Laos – most of which are registered close to the home of Vixay Keosavang, a man involved in wildlife trafficking[19] – they found no trace that the organisations even existed. They were shell companies, most linked to the Xaysavang Network.

According to the 'The Extinction Business' report, South Africa's officially sanctioned quota of lion bones was in fact feeding a murky criminal underworld. 'It is widely accepted that trade on the southeast Asian side is not transparent, properly understood or identified and is associated with wild animal trafficking, poaching and the demise of tigers,' it said. 'Given this knowledge, this highlights a particularly worrying trend and literally means that governments on the supply and demand side, and the CITES Secretariat by implication may be … aiding and abetting criminal activities.'

In 2017, a *Guardian* report by Nick Davies and Oliver Holmes noted that in a single year, 2014, Vinasakhone Trading was authorised by the Laos government to traffic US$16.9 million of animal products through Laos. This included 20 tonnes of ivory, valued at US$5 million. This constitutes a breach of international law. The quota included 10 tonnes of lion bone and 1,300 tonnes of live turtles, snakes, lizards and pangolins, all of it in breach of CITES quotas.[20]

They also reported that Keosavang had been granted permission to trade in 12 different animal species, including monkeys, crocodiles and pangolins; 250 tonnes of soft-shelled turtles (which would lead to the death of around 45,000 turtles); 100,000 python skins; 1,000 magpies; 100 tonnes of dogs (commonly cooked in Vietnamese restaurants); and 20 tonnes of animal bones, presumably from lions.[21]

'The Extinction Business' report warned that the link to organised crime should inform policy decisions around the legal trade. The South African government, it said, was clearly placing the greed and profits of a marginal and problematic grouping before sound and ethical conservation management.[22]

Who was this 'marginal grouping'? It was a handful of brokers and taxidermists who have dominated the lion bone trade for years. Sandra Linde Taxidermy exported more than 1,150 lion skeletons and carcasses between 2016 and 2019, making it the top bone trader during that period and was named – following a question in Parliament – as supplying the Xaysavang Network.

Others supplying Xaysavang, according to 'The Extinction Business' report, were Sebastian Rothmann, Marnus Steyl and Kobus van der Westhuizen. Also exporting bones were JP Wapenaar, Hatari Taxidermy, GJ van Zyl and T Cloete.[23]

Key farms supplying bones listed by forensic investigator Paul O'Sullivan were Shangwari Safaris, Steyl Game CC and Leeuwbosch Game Lodge.[24] Importers, all Laos-registered, were S Durosagham, Sipharpra Duarseram, Vixay Keosovang, Jacek Raczka and Bounpasong Paphatsalang.

Keosavang's role in widespread illegal wild animal trade had been noted by various law enforcement agencies in Asia from as early as 2003. In 2013, the US government offered a US$1 million reward for information leading to the dismantling of his organisation. Despite widespread knowledge that the Xasavang Network was a criminal syndicate, South Africa continued to permit the provision of CITES export permits.

The head of Asia Investigations for the Freeland Foundation, Steve Galster, noted that 'Watching Vixay [Keosavang] operate for many years made me realise that the biggest threat to wildlife is the legal wildlife trade.'[25]

Barbara Creecy moves the dial

Following the death of Jabula in 2018 and the discovery of the Wag-'n-Bietjie and other lion abattoirs, major big cat conservation groups such as the Endangered Wildlife Trust (EWT), Panthera, the EMS Foundation, Ban Animal Trading, Wildlife ACT, WILDTRUST, Blood Lions and the National Association of Conservancies called on South Africa's then Minister of Environmental Affairs, Dr Edna Molewa, to regulate the captive breeding carnivore industry to avoid welfare abuses.

The evidence in 'The Extinction Business' investigation was hard to ignore. It traced lion bones from the farm gate to consumers in Laos, Vietnam and Thailand; its cover picture was of the doomed Jabula staring bleakly out of a barred cage shortly before his death.[26] Slowly, official wheels began to turn in the right direction for beleaguered lions.

In 2015, the Department of Environmental Affairs had issued a Biodiversity Management Plan for the African Lion (*Panthera leo*), which proposed a plan to ensure 'viable and ecologically functional populations of managed and wild lions, along with well-managed captive populations'. While still embracing captive breeding, it called for the establishment of a lion working group to plan the way forward.[27]

Then, in August 2018, the Department of Environmental Affairs took the first steps towards regulation. A two-day colloquium on captive lion breeding was called by the Portfolio Committee on Environmental Affairs. Hearing views from scientists, economists, breeders, hunters and conservation NGOs, it concluded that captive lion breeding and canned hunting had done serious damage to South Africa's international image and could show no conservation value.[28]

Its chairman, Mohlopi Mapulane, said, 'We need to come up with a solution as quickly as we can as we can't allow for our international reputation to be soiled.' He added there was no scientific or conservation proof that the lion bone trade had any tourism advantage. 'Even the hunting organisations are turning their backs on the industry. We must do something as a country,' he said and questioned how quotas for hunting and bone export were determined.

Key outcomes of the colloquium were:

- There is no conservation benefit to captive breeding of lions[29] or the lion bone trade;[30]
- The captive lion industry is damaging to South Africa's image as a conservation leader and poses a substantial risk to our tourism industry;[31]
- The use of lion bones and body parts and derivatives for traditional medicine is one of the major threats to wild lions and could serve as a cover for illegally wild-sourced lion and other big cat parts.[32]

There's a general abhorrence of the industry across multiple sectors including animal welfare, animal rights, conservation and hunting organisations, which echo the South African public sentiment.[33]

The National Assembly adopted the resolutions of the colloquium, which required the Department of Environmental Affairs 'as a matter of urgency' to initiate policy and a legislative review of the captive breeding of lions and the bone trade and submit quarterly reports to the Portfolio Committee on its progress.

The following year, new Minister of Forestry, Fishers and the Environment, Barbara Creecy announced the formation of a high-level panel to review the existing policies, legislation and practices related to the management, breeding, hunting, trade and handling of elephant, lion, leopard and rhinoceros. However, disappointingly, provincial authorities reissued permits for 88 lion-breeding facilities.[34]

The panel noted major contraventions in the breeding industry and signalled a powerful new dispensation for wild animals; and it appeared to mark the beginning of the end for captive lion breeding. Its majority view was that 'South Africa will not captive breed lions, keep lions in captivity or use captive lions or their derivatives commercially.'

A large commercial captive lion breeding facility near Wolmaransstad, South Africa.

In July the following year, Creecy gazetted a draft 'White Paper on the Conservation and Sustainable Use of South Africa's Biodiversity'.[35] Its definition of sustainable use was a shot across the bows of those who used it as justification for wildlife farming and trophy hunting and one of the most progressive in the world.

Species could be 'used' only as long as the ecological integrity of the ecosystem in which they lived was not disrupted and their wellbeing was humane and not compromised. According to the White Paper, conditions under which an animal was kept had to 'be conducive to its physical, psychological and mental health and quality of life, including its ability to cope with its life'.

In 2022, a task team was appointed to recommend voluntary exit options and pathways for the captive lion industry. Its mission is to do an audit of all breeding facilities and plan the dismantling of the industry.

The following year the Department of Forestry, Fisheries and the Environment called for public comment on its intention to block future lion breeders and prevent the exploitation of a long list of animals, birds, insects, fish and plants. The same year, the department issued another White Paper on the sustainable use of biodiversity, which further defined the notions of wellbeing and sustainable use of wildlife. A specialist panel of experts was then formed to propose steps for the closure of captive breeding facilities and a notice was issued seeking information on voluntary closures. The outcome was tabled in Parliament in April 2024.

This pause coincided with Cabinet's approval of a Policy Position to end the captive breeding of lions and rhinos, the closure of captive lion facilities and the end of 'canned' hunts. The acceptance of the Policy Position coincided with the release for discussion by the Department of Forestry, Fisheries and the Environment's National Biodiversity Economy Strategy. This proposed to grow areas under conservation – called 'mega living conservation landscapes' – from 20 million hectares to 34 million hectares by 2040, an area equal to seven Kruger National Parks.

The strategy, while proposing wide areas of the country under conservation, caused unease among conservationists over its implementation under the banner of sustainable use. In 2019, more than 30 wild animals, including lion, giraffe, white and black rhino and cheetah, were listed under the Animal Improvement Act, effectively rendering them farm animals subject to manipulation and consumption.[36] Shortly afterwards, 98 more wild animals were proposed to be listed under the Meat Safety Act, including rhino, hippo, elephant and crocodile.[37] According to the Act, they could then be 'slaughtered for food for human and animal consumption'.

It has been a long and complicated journey, but the overall direction has been positive. However, implementation will not be easy. There is a likelihood of endless litigation from lion breeders. Lion farming is governed by a patchwork of contrasting legislation across multiple provincial and national authorities, with disparities and legal loopholes that create opportunities for harmful and fraudulent activity. In the absence of a full national audit, the scope and scale of the commercial captive–predator industry is largely unknown and publicly available information is lacking.[38]

Bones of contention ~ 429

SOUTH AFRICA

42

Killing the king

Breeding, killing and exploiting captive-bred lions for trophies and other commercial reasons has nothing to do with conservation or education and everything to do with a society's moral and ethical compass.

Ian Michler

In most societies today, humans have a legal right to life, whereas non-human species do not. This discrepancy is why humans can be cruel to animals with little consequence. This explanation is straightforward, however, and deeper examination of human-to-animal behaviour reveals a far more complex set of relationships that exists across various spheres.

It is important to acknowledge that not all human behaviour towards animals is abusive or disrespectful. There is clear affection between pet owners and their animals, and many individuals and communities still revere their spiritual and totem species, often expressing this through recreational, cultural and artistic means. Furthermore, there are numerous animal lovers involved in conservation, welfare and science, all clearly concerned for the individuals and groups within their care.

Conversely, within society there may be a spectrum of animal abuse, ranging from individual acts of cruelty, such as someone kicking their pet or beating a hobbled donkey in a field, to organised activities such as horse racing, bullfighting and dog fighting. Additionally, there is the deeply entrenched exploitation of animals that occurs in food production systems. Human benefit is sometimes used as a justification for animal exploitation. In South Africa, for example, the use of wild animals is frequently viewed from a utilitarian perspective, where human benefit is deemed acceptable, and sustainable use makes it further justifiable.

Investigations and blood lions

These musings have been uppermost in my mind since those shocking forays over 25 years ago when I first set eyes on countless caged lions. If ever there is behaviour reflecting the murkier side of humanity's mindset towards the natural environment and wild species, the intensive breeding of thousands upon thousands of predators ranks among them. To view these industries up close is to look into a dark prism, one that scatters insight into a spectrum of perverse and offensive human attitudes towards animals.

It was the scale and brutality of the agriculturalised settings that remain so disturbing to this day. Rows

of large concrete and wire enclosures, mounds of excrement and chewed bones, filthy drinking troughs and the mindless up-and-down pacing of a few that told of the frenzy to escape. Bereft of their wildness, the rest seemed emaciated, lacking the vigour to move. Without cover, they just lay panting under the baking midday heat or sprawled across concrete floors.

Seeing lions and other species under these conditions was an awful sucker punch to everything I knew and understood about the natural world as well as ecological systems, not to mention the poetry of lions, the most charismatic of species that roam Africa's plains. I returned from that trip compelled to contribute towards ending the predator-breeding and canned-hunting industries.

On this continent, South Africa lies at the heart of these practices, but Namibia, Botswana, Zimbabwe, Zambia and, more recently, Tanzania are also involved. The vast majority of animals churned out in breeding facilities are lions, but species such as cheetahs as well as a host of exotics, particularly tigers, are included.

Surprisingly, the early expositions received little traction and, despite the first-hand accounts, few seemed to believe what was happening behind the high fences. This was partly thanks to the law; keeping and killing lions was not illegal. However, various aspects of the industry – such as the widespread use of veterinary drugs, the movement of animals across provincial boundaries and welfare violations – were clear transgressions and much of what was put forward as justification seemed to be misinformation and at odds with accepted conservation principles.

It certainly didn't help that, other than the International Fund for Animal Welfare (IFAW), every major conservation agency in South Africa chose to stay away from an issue deemed too controversial to handle. Avoiding the issue created a vacuum that allowed the industries to operate with little scrutiny.

Along with these legal screens and the ignorance and blind eyes turned to both the conservation racket and any ethical or moral considerations, compliant provincial authorities served to boost the breeders and canned hunters. There was an attempt by the national government in the mid-2000s to muzzle the industry, but they bungled it, losing a critical court case in 2010. This legal loss served to further stimulate growth. With no appeal or additional action taken by government, it served as a green light to operators.

It was only once the award-winning documentary film *Blood Lions* was released that people got to see and understand the horrors and extent of the industry. The visual impact of lions being farmed and killed plus an endless stream of cubs used as playthings in tourism facilities began grabbing the world's attention. The global campaign to counter the myths and misinformation also kicked in.

But as peculiar as this may seem, after 25 years of work by so many, we remain none the wiser as to the exact number of predators or facilities that hold them. My first guesstimate was approximately 800 animals, and that was back in the late 1990s. In 2005, I reported a range of 3,000 to 3,500 but, by the time we had completed research for *Blood Lions*, it was up to between 6,000 and 8,000 predators in over 200 captive facilities. When Lord Michael Ashcroft released his book *Unfair Game* in 2020, he reported around 10,000 to 12,000 animals in over 300 facilities. Asking national or provincial authorities – the sources one would typically expect to have the answers – has never been a reliable option as they do not know either.

By way of interest, South Africa and its neighbours are not alone in treating predators in this manner. According to IFAW, the USA has about 10,000 predators in captivity, the majority being tigers, and there is a robust trade in various species within numerous Gulf states as well as in parts of Asia.

Killing the king

It soon became apparent this entire industry was based on one simple goal: profit. Initially, it was mostly about trading live animals and canned hunting, a shameful practice of killing lions (and other species) in small, enclosed areas. At times, with the 'prey' baited or drugged, the so-called hunter could bag a guaranteed trophy without paying the significant sums charged for hunting lions in the wild.

In the early years, the operators seldom tried to explain their actions. It was simply hunting and a way to earn a decent living. It was only once the media got hold of the more shocking details that the justifications attempting to link the practices to conservation and science appeared.

Like any good business model, the breeders and operators started expanding their range of products. This included the insidious interactive tourism products: petting lion cubs, walking with lions, and filming and photographic opportunities that relied on a steady supply of predators to fill their facilities. More recently, the trade in lion bones and body parts spread rapidly. To meet the demand for animals, breeding and trading grew exponentially.

Over the years, I have noticed three principal themes pertaining to the nature of these industries and those involved. First, there are the religious convictions of those that believe humans have dominion over all other species. For them, this makes the use and abuse of lions perfectly acceptable. And, while filming *Blood Lions*, we had hunters claiming their faith somehow sanctioned the killing of animals for trophies.

The second theme relates to a lack of consideration towards animal welfare, let alone the subject of animal rights. In essence, so many of the early operators came out of the apartheid era, a brutal system based on domination and discrimination framed by denying the majority their human rights. With this background, it was hardly surprising to find little recognition of animals' sentience or of the need to treat them with respect. As any psychologist will tell you, when we diminish or belittle the capacity of others, the chances of recourse to ill-treatment and abuse are high.

And lastly, there's the cold commercial approach, one that relies on the doctrine of sustainable use. It is fuelled by a small group of consultants, economists and lobbyists who have managed to extract and interpret a harsh and self-serving utilitarian interpretation of sustainability from wording in South Africa's constitution.

This thinking attempts to validate the activities on the basis of claims that they provide economic

White lion cubs that have been bred for the lion petting and hunting industry.

opportunities and that breeding large numbers of lions under captive conditions brought conservation benefits for the species. For the most part, this last argument was bought by the governments of the day as well as by some within the wider conservation and tourism circles.

The case against

I don't know how many times this needs to be said, but other than the South African government (until very recently that is) and the operators involved, most of research, opinion and public sentiment has been against predator breeding, canned hunting and interactive tourism products. This opposition has been voiced repeatedly in many forums and platforms in South Africa and across the world.

The recognised scientific community unanimously holds there to be no conservation benefit, a position reinforced by the fact that captive-bred lions are never included in populations considered for conservation purposes. The gene pool has been compromised, the lions lack the vigour that comes with wildness, and they have lost their fear of humans, making them dangerous in situations involving handlers and tourists.

An additional conservation concern relates to lion-breeding farms and a legal lion-bone trade becoming channels promoting the poaching of wild lions and an illegal trade in body parts. This, in turn, could threaten the survival of wild lion populations.

The educational claims are wide of the mark. Everything about these facilities is contrary to the life of lions in the wild, and 20 to 30 minutes of petting cubs, walking with subadults or getting close-up images simply reinforces the cycle of abuse. In fact, these exhibits offer nothing more than crude entertainment. In addition, experiencing lions under these conditions is the antithesis of lavish marketing claims made by South African tourism authorities of the country's destinations being ethical, responsible and authentic.

The industry is not sustainable. The breeding, based on human selection over natural selection, leads to domestication. Once the canned hunters and ignorant tourists realise lions have lost their wildness, the thrill of killing or petting a tame pussy cat will disappear. Those peddling the products will be left with selling the hype and a lie, with thousands of compromised lions unable to fend for themselves.

Even the wider hunting industry across the world has distanced itself. Some bodies have banned captive breeders and hunters from displaying and marketing at hunting fairs, while in South Africa, the local hunting association has split. The majority left the original organisation to start a new one, this time based on principles that object to canned or captive lion hunting.

More recently, the industry has been trying to pin its continued existence on economic arguments. Their interests represent a fraction of the much larger tourism pie and, as economic reviews have highlighted, global boycotts of the country's entire tourism industry by opponents will have significantly greater negative impacts.

More specifically, the claims the industry makes about being a job creator are likely negated by the volunteer tourism sector. Farms and facilities bring in dozens of paying volunteers, most from outside the country, as another source

> **The only solace in this worst-case scenario is that the scientific and conservation community now has a clear historical record concluding that breeding lions and other predators in captivity to be killed or abused for commercial purposes has no conservation or educational value.**

of revenue. In exchange for being able to interact with and cuddle lions, they pay the operator for the opportunity to live and work on these farms and tourism facilities. It is an arrangement that may well be a job destroyer as the volunteers deny local South Africans work.

There have also been puzzling claims about the industry being a vehicle for transformation. Given the historical and political backgrounds of so many within the sector, it's highly unlikely it can be a platform for addressing transformative rural development of any sort. When all these counterarguments are measured against the risks the industry poses to South Africa's tourism and conservation sectors, the case for predator-breeding and its related activities makes no sense at all.

Working with hope

Despite the fact that these industries continue to survive, there is hope. In this regard, Sunday, 2 May 2021 has gone down as a notable day for captive lions and those opposing the industries. On that day, Barbara Creecy, South Africa's Minister of Forestry, Fisheries and the Environment, spoke the words so many of us feared we would never hear. 'The panel recommends that South Africa does not captive-breed lions, keep lions in captivity, or use captive lions or their derivatives commercially,' she stated. 'I have requested the department to action this accordingly and ensure that the necessary consultation in implementation is conducted.'[1]

Her announcement came after a lengthy consultation process, conducted through a government-appointed High-Level Panel. This panel heard countless presentations and voices across the entire spectrum of stakeholders, including those involved in lion breeding, and the majority voted clearly and unambiguously to end the abuse and commercialisation of captive predators.

For so many, her speech was a victory of sorts. Despite previous setbacks and the grey areas in the report, it heralded a significant shift in thinking.

For the first time, the principles of science, conservation, ethical tourism and animal welfare trumped the powerful commercial lobby of those profiting from the exploitation.

However, there was a sobering component. These were merely recommendations, and another monumental effort would be required to see them through the legislative process. For *Blood Lions* and the broad church that supports the work of the minister and the panel, it continues to be about holding the majority bloc together while supporting the government with the expertise needed to ensure they complete the work.

Ending the commercialisation of Africa's most iconic species will carry a softer message, one that speaks of an ethical reckoning. Breeding lions to exploit them is simply another example of the dysfunctional relationship humanity currently has with the natural world. How else does one explain our continually expressed love for animals and concerns for conservation translating into the abuse we impose on them and the relentless, if inadvertent, drive to extermination? Ending these industries will say something about restoring a semblance of truth to claims of being a progressive and humane society.

As a footnote, it is worth pointing out that, while the process towards closure brings hope, we must be realistic. It's not beyond the realms of South Africa's political reality that corruption, political manipulation and expediency could well still derail what is almost complete. Clearly, there are already signs of foot-dragging as provincial departments have failed to implement any of the recommendations of the High-Level Panel or the minister. If this does happen, it would be an awful indictment on those involved, and another shocking example of leadership failure, underpinned by a disregard for non-human animals.

The only solace in this worst-case scenario is that the scientific and conservation community now has a clear historical record concluding that breeding lions and other predators in captivity to be killed or abused for commercial purposes has no conservation or educational value.

Lion carcasses from legal hunts, hanging up to dry on a hunting concession in the North West province, South Africa. With fewer than 3,000 tigers left in the wild, lion bone is becoming an increasingly popular substitute in Asian markets. The bones are used to make 'tiger' wine and are crushed and used in Asian medicines. © Brent Stirton

PORTFOLIO

Brent Stirton

Most people feel there's a certain grandeur to lions and a genuine dignity. As apex predators, lions have earned the mantle of 'king of the beasts'. Canned lion hunting is the kind of trophy hunting that fails to consider this at all: there's no respect, no reverence.

It's a sad experience to walk through the breeding areas and see long lines of caged enclosures, condemned lions on all sides, sitting around, unable to be lions. Chicken feathers adorn the fences keeping them in. The lions lack purpose, they are wondering what they're doing there, every gene in their bodies reminding them they were meant for the bush. Some pace endlessly, others lie in groups close to the fence. Yet others are fierce and roar at you. Every hair on your body stands up: these are still lions, kept at bay only by a very thin fence.

Hunting lions this way is about wealthy faux thrill-seekers ticking a box. In contrast, I've been on a couple of fair-chase hunts. One went on for 10 days and, in the end, the lion evaded the hunter despite the professional hunter's (PHs) doing everything they could to entice the lion into a killing position. The hunter on that one was philosophical: 'He was smarter than us, he deserved to get away.'

Compare this to a small enclosure, where the lion has no means of escape. On the two canned lion hunts I covered, in both instances the lions were obviously still lethargic from the transport tranquilliser and had not fully recovered from being taken from the breeding cage to the hunting zone. One could barely raise her head, the other had to be antagonised before it reluctantly stood up to be shot.

The PHs' main responsibility seemed to be to create hype and excitement where there was none. This was not a deadly situation, there was zero imminent threat – and, in the unlikely event of real risk, the two PHs stood ready to protect the client. I watched an overweight American stumble forward with his bow and shoot the lioness, missing a kill shot the first time and finishing it with the second. The lion never moved in his direction at all.

My first thought was how underwhelming and pathetic it seemed. I'm not against hunting, maybe it has its place, but this was something else. The frenzied commentary of the PHs gave the impression that the bow-hunter had saved us all from a certain encounter with death. But canned lion hunting is only the illusion of a real lion hunt, a brutal convenience for those who want the acclaim and the trophy without the drudgery, the real danger or the risk that the lion could give them the slip. Canned hunters just don't understand the honour in fair chase.

Wandering away from the lion enclosures, I stumbled upon an apocalyptic scene. I feel as though I was led there. Tucked into a corner of a construction site were at least 40 lion skeletons drying on a fence. It took me a little while to comprehend what I was seeing. As I tentatively took my first photos of this carnage, a man walked up carrying another skeleton, meat scraps still pink on the bone. He hung it up with the others, one of the hind legs sticking out at a strange angle. The man smiled at me on his way out, commented on the heat, and went about his day.

Most hunters take the skull and skin but very few take the bones. These make their way to Asia, most often to masquerade as tiger bone for traditional Asian medicine or wine. This is the hidden aspect of the canned lion industry and the guarantee for wily breeders that their return on investment is assured.

CITES issues permits for these bones to be exported but the South African authorities have put a hold on this – for now.

> In 2020, the South African government made noises about shutting down this industry. There are well over 10,000 lions in the canned industry; the question is – what to do with all of them?

439

Top and above: An American bow hunter hunts a lioness with professional guides on a game farm close to the South Africa/Botswana border. The lioness in this image was raised at a captive lion breeding facility and released into the reserve 96 hours ahead of the hunt.

Left: Breeding cages on a lion breeding farm outside of Wolmaransstad, South Africa. These lions are born in these facilities and are raised to maturity in cages. When they are old enough (7 years is the preferred age), they will be sold for lion hunts anywhere in the country. As controversial as the practice is, it is legal under the current South African judicial system.

An aerial view of the largest lion breeding facility in the world, located near Koster, a small town around 100 miles to the northwest of Johannesburg, South Africa. At its peak, this single farm housed between 500 and 600 lions, the majority of which were bred for the trophy hunting industry. © Brent Stirton

A lioness is killed by a herd of buffalo in Kruger National Park, South Africa. © Kyle Mills

Kruger National Park – the end of a predator

Kyle Mills

The following drama occurred in the late afternoon at Mestel dam in the Kruger National Park. I arrived at the dam around 3:30pm as a stop on an afternoon drive I had planned. I ended up sighting a leopard across the water and so decided to stay for a while as it began (unsuccessfully) stalking some impala.

Just before 5pm a large herd of buffalo came down to drink and chased the leopard up a tree. At this point some of the herd split off, walking along the water's edge when, seemingly for no reason, the rest of the herd began running in that direction as well.

I began shooting the scene, purely for the aesthetic of the dust in the afternoon light when I noticed a commotion in the midst of the herd. I could just detect a lioness in the fracas, and that it was under attack. I kept shooting.

The buffalo were ruthless in their attack and didn't stop until it was evident that the lioness was dead. Then they started to wander off. There was a second lioness that tentatively confronted the stragglers of the herd, but her heart wasn't in it. Once the herd of buffalo had gone, she approached the body of her companion, sniffing at it before moving away slightly and lying down.

I sat watching as long as I could but had to leave in order to get back to camp before the gates closed. The next morning I went back and only the body was there.

No other pride members arrived throughout the ordeal – just the two lionesses up against a large herd of buffalo. The lionesses most likely had cubs waiting to be fed. Or perhaps they weren't hunting and ended up just being in the wrong place at the wrong time.

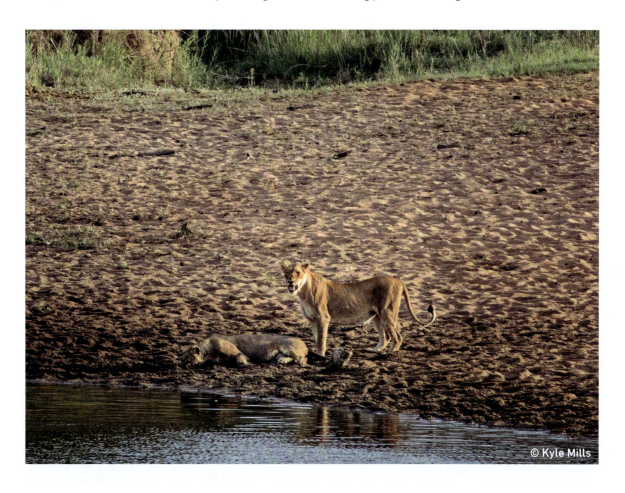

INDIA

43

The last lions of Asia

Long, long ago, lions from Africa dispersed into other regions and made their homes in lands far from where they began. Most of those early travellers have disappeared but, in the Gir Forest in India, they live on, protected and revered.

Bhushan Pandya

Most people associate lions with the African continent. However, in the past, subspecies of African lions roamed many countries in Europe and Asia, too. Trophy hunting, loss of habitat, decreasing density of large prey ungulates, deforestation and land use changes have progressively wiped out these populations almost everywhere – except in the Gir Forest of India.

Here, however, questions arise that are very different to those you'd ask in Africa. Lions are known to live in the savanna: can they survive in the dense and thorny indigenous vegetation, or Prosopis, of coastal areas? Can a top predator be adored by humans, like a family member? Do villagers host a *besna* (post-death prayer ritual) for a wild lion? Can an apex carnivore be worshipped as a god? Do some communities adopt the title of *sinh* or *singh* (lion) while naming their sons? Is the Asiatic or Persian lion, *Panthera leo persica*, the same as African lion subspecies, *Panthera leo leo*?

All these seemingly absurd questions have one answer – YES! You would have to visit Gir Forest, situated in the Saurashtra region of Gujarat state in western India, to see for yourself. This part of Gujarat, formerly known as Kathiawar, has Asia's largest compact, dry deciduous forest. Gir National Park, Gir Wildlife Sanctuary, Pania Wildlife Sanctuary, Mitiyala Wildlife Sanctuary and peripheral reserves and protected forest make up what is collectively known as Gir Forest.

Identifying Asiatic lions

Asiatic lions have shorter and sparser manes, slightly smaller body size, a longitudinal belly-fold (loose skin on the abdomen), and larger elbow and tail tufts when compared to their African cousins. They also have two bifurcated infraorbital foramina (small holes in the skull) while African lions have one foramen.

In terms of an earlier classification, there were 11 subspecies of lions in the world. However, after in-depth genome sequencing performed by 23 experts of the IUCN Species Survival Commission Cat Classification Task Force, the old taxonomy has been revised.[1] The classification of species can no longer be based merely on morphological characteristics like belly-fold, size of

Gir Forest, India. © Brent Stirton

mane, etc. In terms of the new classification, there are only two subspecies of lions: *Panthera leo leo* and *Panthera leo melanochaita*.

Panthera leo persica is now reclassified as *Panthera leo leo*; they are believed to have come to Asia from western and Central Africa. However, officially, the Asiatic lion is still *Panthera leo persica*. In terms of genetic discrepancy, however, scientific studies have revealed that Gir lions are genetically as diverse as Royal Bengal tigers.

History of Asia's lions

The British officers and Indian maharajas of pre-independent India used to regularly hunt lions (and tigers) for trophies. Some of the British officers' hunting activity almost wiped out the lions from large areas. Colonel George Acland Smith is said to have killed more than 300 lions in 1857–1858. William Fraser shot 84 lions. William Rice shot 14 lions in Gir on a single shooting excursion. During the 50-year period from 1820 to 1870, 1,500 lions are believed to have been hunted. From 1875 to 1925, over 80,000 tigers, 150,000 leopards and 200,000 wolves were slaughtered.[2]

The *Kathiawar Gazetteer* reported in 1884 that there were 'probably no more than ten or a dozen lions left in the whole Gir Forest'. In 1893, Junagadh state reported a count of 31 lions. However, both the estimates were 'misleading' and 'incorrect' according to many experts, who believe the lion population never fell below 60 to 100 at any time. These varying figures were arrived at thanks to the limitations of the census methodology and the vast area covered. At the time, there were lions in the other princely states, Baroda and Bhavnagar also.

The beginning of conservation

A significant improvement in lion protection came about when Lord Curzon, the British Viceroy, was invited by the eighth Nawab of Junagadh, H H Muhammad Rasul Khanji Babi, for *shikar* (hunting) in 1900. A letter written by a concerned citizen was published in a Bombay daily newspaper titled 'Viceroy or Vandal?', criticising the proposed hunting expedition. Moved by the report, the Viceroy cancelled the hunting excursion and urged the Nawab to enforce strict protection. The Nawab was displeased and later wrote to the Viceroy, insisting the lion population was not at risk. In spite of this, he responded to the Viceroy's suggestions and implemented laws to protect lions in Junagadh state.[3]

After this point, permission was needed to hunt a lion. However, it was given rather liberally to British officers. Also, the rulers of the two adjoining states, namely Baroda and Bhavnagar, formed policies but did not strictly enforce them. There were incidents of luring lions out of protected areas using live bait. Later, lions were hunted by 'sportsmen' despite the difficulty in getting permission from the Junagadh state administration. When Nawab Rasul Khanji died in 1911, his successors, including administrator H D Rendall Esq. (1911–1920) and later, Muhammad Mahabat Khanji III (1920–1948), the last Nawab of Junagadh, retained his policy of protection.

On 15 August 1947, 562 princely states merged to become one nation, India. Since then, consistent and collective efforts by both state and central governments, media, scientists, NGOs and local people have scripted a rare success story of Asiatic lion conservation. By this time, though, no lions remained outside of the Gir area and, for that reason, the Asiatic lion is also popularly known as the Gir lion.

Gir Wildlife Sanctuary was declared on 18 September 1965, making it one of the oldest sanctuaries of the country. Its core area of 258.7km² was given the highest protection when it was declared the Gir National Park in 1975, and the total area of 1,412km² was brought under protection.

Wild Life (Protection) Act, 1972

Dr M K Ranjitsinh, IAS (Retd.), hailing from the royal family of a princely state, Wankaner in present-day Gujarat, suggested to then Prime Minister Indira Gandhi that special laws were needed to protect the forests and wildlife. This resulted in the Wild Life (Protection) Act, 1972[4], drafted by Dr Ranjitsinh.

With the support of the Act, the population of wild ungulates such as chital (spotted deer), chinkara (Indian gazelle), sambar (the largest antler in Asia), nilgai (blue bull) and wild pig started increasing. The spotted deer population in Gir has increased from just a few thousand in the 1970s to 88,328 according to the last wild prey species population study in 2024.[5] Other prey species' populations have similarly grown multifold. As a result, the prey preference of lions inside the Gir National Park and Wildlife Sanctuary has changed. Their diet used to consist of 75% domestic cattle and 25% wild herbivores. The latest scat analysis shows 74% wild herbivores and 26% domestic livestock. Sambar and spotted deer are the preferred prey species, forming 35% and 19% of lions' diets, respectively.[6] Overall, in Gir Forest, the population of wild prey species has increased from 167,579 in 2019 to 219,021 in 2024.

The Maldharis (pastoralists)

For decades, the Maldharis and Siddis (Afro-descended people of Gir) have co-existed with lions and leopards without much direct conflict. However, the area of Gir is limited. Both the Maldharis and wildlife depend on common natural resources like land, water and grass. There are still 44 Maldhari *nesses* (temporary human settlements) within the Gir Wildlife Sanctuary and their families have been expanding. Efforts to relocate people out of the protected area have not been entirely successful. Many relocated Maldharis have illegally returned, nullifying the efforts to make more inviolate space for lions. Scientists and park managers agree that these families should be shifted elsewhere, but that the process would have to be systematic and gradual.

The shifting of Maldhari families has been made more complex because of the Forest Rights Act. Some of the Maldharis demand facilities like tar roads, electricity, buildings, schools and hospitals inside the sanctuary, developments that would pose a threat to wildlife. In 1982, the government of Gujarat showed the intention to declare the whole of Gir Protected Area a National Park and it seems the time has come to implement this intention. The Maldhari families could be encouraged to move with the offer of an acceptable package and sensible counselling. Their translocation out of the Gir would greatly benefit their future as well as that of the lions.

Expanding the range

The growing wild ungulate population – blue bulls and wild pigs in particular – has spread out of the protected areas and, since the early 1980s, has been followed by lions and leopards. After Gir, the state authorities declared Barda (1979), Rampara (1988), Pania (1989), Mitiyala (2004) and Girnar (2008) as wildlife sanctuaries. At present, the Asiatic Lion Landscape (ALL)[7] or the lions' home range has expanded to about 30,000km^2 in and around more than 1,500 villages.

The lion home range outside these protected areas is known as Greater Gir and the lion population within this range has been increasing, mainly because the villagers, and particularly farmers, have accepted lions into their lives. In this large area, the lions use grass *vidis* (grasslands), degraded land, coastal areas, stretches along the Shetrunji River banks and farm fields to roam and live. Most of the farmers are happy with the situation since their precious crops are protected from wild ungulates like the blue bulls and wild pigs when lions are around.

Developing gene pools

Barda Wildlife Sanctuary near Porbandar and Rampara Wildlife Sanctuary near Rajkot have (captive but purebred) Asiatic lions. These sanctuaries are being prepared to accommodate free-ranging wild lions in the near future. The area of Barda Wildlife Sanctuary is 192.3km^2, slightly larger than that of Girnar Wildlife Sanctuary. The surrounding area of more than 350km^2 is also suitable for about 125 lions.

A three-and-a-half-year-old male reached Barda in January 2023, becoming the first lion to have settled there since 1879 and two females have been released to augment the lion population there. From being merely a sanctuary, it has become a second home range for lions within Gujarat state.

Gir Forest, India. © Bhushan Pandya

My experience is that the more I know about wild animals, the more I fear humans. We have turned our back on nature. It is time we go back to nature.

Lion Translocation Project

After a hard-fought legal battle, the Supreme Court of India gave a verdict on 15 April 2013 in favour of shifting lions to Palpur-Kuno Wildlife Sanctuary, Madhya Pradesh.[8] An expert committee was formed to oversee the move in accordance with the guidelines of the International Union for Conservation of Nature (IUCN). However, since September 2022, another relocation programme has enjoyed priority: a consignment of 20 African cheetahs sourced from Namibia and South Africa have been brought to Palpur-Kuno under a five-year cheetah reintroduction project. This equally ambitious operation has placed the Lion Translocation Project on hold.

Risks in Greater Gir

'Protecting the protected species in protected areas is rather easy. However, protecting the protected species outside the protected areas is a challenge,' according to conservationist B J Pathak, IFS (Retd.).[9] There are multiple threats to lions from, for example, speeding vehicles on state and national highways, high-speed trains, open wells in farms and the illegal electrification of farm fences. The government of Gujarat has taken several precautionary measures to avoid unfortunate incidents. Yet, looking at the large areas concerned, the burgeoning human presence and the ongoing development of infrastructure, the spread of lions into unprotected areas is bound to lead to problems.

Recently a vision document prepared by the Gujarat Forest Department was developed.[10] It is an ambitious roadmap detailing lion protection, conservation and management for the next 25 years. The document calls for ₹2,927 crore (about US$352 million) for activities such as infrastructure development, national wildlife disease diagnosis, a research and referral centre, habitat improvement, covering open wells with parapet walls, retrofitting and proactive mitigation measures on wildlife corridors for the next 10 years. Furthermore, a state-of-the-art National Wildlife Disease Diagnostic, Research and Referral Centre will be built near Junagadh.

Ecological significance

Gir has seven perennial rivers and four dams. They are the lifelines of many in this semi-arid landscape. The Gir Forest has an undulating terrain of volcanic hills that haven't seen much human intrusion such as industrialisation, cultivation or mining. The hilly corridors are protected by nature, allowing lions and other wildlife to disperse. The Gir Forest is bordered by the sea on three sides, the nearest being about 40km away. The forest cover restricts the salinity ingress by recharging the potable groundwater, thus providing an invaluable service to the ecosystem.

In addition to the source population of Asiatic lions at Gir, Pania, and adjoining areas, according to the latest estimate (June 2020), there are seven satellite populations.

Natural calamities

In September 2018, 23 lions died in four weeks in the Dalkhaniya range of eastern Gir from the canine distemper virus and babesiosis. An ambitious emergency rescue operation was carried out by the management authorities and the spread was contained. In April 2020, there was some spread of babesiosis and probably an undetected virus. Many lions were rescued, treated and released back into the wild.

On 17 May 2021, Tauktae, a cyclone with torrential rain and wind speeds of 165km/h, struck the Asiatic Lion Landscape. Over 3.5 million trees were uprooted and several hundred nesting and roosting common birds lost their lives.[11] All the lions were found to be safe after the cyclone. While it is not possible to protect trees from cyclones, nature has opened up quite a few areas in its own way, through the natural thinning of dense habitats. Moreover, if we go back in history and see the incidence of cyclone occurrence in the landscape, we realise there has been a major cyclone every 40–50 years, which leads to the uprooting of millions of trees and opens up the areas. In the long run, by such processes, nature takes its own course through vegetational succession and gains the momentum to provide a good habitat for wildlife.

Lion census and Poonam Avlokan

The lion census is an important event spanning around six days and occurring every five years in the summertime. The population estimate is done by the Direct Beat Verification or Block Count method.[12] Proper documentation (filling out special forms, recording GPS readings, photography and videography) is undertaken by trained teams in the presence of various stakeholders, including scientists, NGOs and wildlife experts. The final figure is the minimum count: only absolutely confirmed figures are accepted, arrived at after a direct headcount on the particular day. Unlike in tiger or leopard censuses, no indirect sighting is considered in the census.

Since 2014, observations of lions are also performed every month involving a 24-hour lion-population estimate and monitoring exercise. This is conducted on a full moon night. In 2020, the results showed a 28.9% increase in the lion population from 523 to 674 and a 36% increase in the home range from 22,000km² to 30,000km².

What can tourists expect in Gir?

The Gir National Park and Wildlife Sanctuary is a large university of nature, offering its diversity of 338 bird species, 47 reptile species, 41 mammal species, over 2,000 species of insect and over 631 species of flora. There is a well-organised system in place for tourists and, given the wealth of wildlife, ample opportunities for visitors to view a range of animals.

Conclusion

We are in the midst of the sixth mass extinction. The previous five were because of natural causes, whereas the present one is largely owing to human activities. Still, with dedicated collective efforts, climate change can be reversed. The IUCN status of the Asiatic lion has been upgraded from Critically Endangered (CR) in the year 2000, to Endangered (EN) in 2008,[13] and the latest global report[14] has reclassified the species to the Vulnerable (VU) category. This rare conservation story offers a hopeful lesson.

A game drive in the Gir Forest, India. © Bhushan Pandya

44

Under-tourism, flight shaming and the scourge of over-tourism

We need to exempt Africa from climate-based flight shaming while making sure we don't overload the more popular tourist destinations.

Colin Bell

For lions to survive and thrive well into the future, they have to have large tracts of pristine interconnected wilderness teeming with wildlife to allow them the freedom to roam, hunt, raise families and flourish. The problem is people pressure. The map opposite shows where the planet's next billion people will be living in just a few decades' time.

That looming level of human overcrowding around most of Africa doesn't leave much space for lion and wildlife populations and will signal the end of free-ranging, far-roaming nomadic lions. Lions will then most likely be found only within the boundaries of the continent's formal national parks, game reserves, wildlife management areas and conservancies. But the COVID years proved that the parks and reserves around Africa become unviable when there are few or no tourists to generate the revenues needed to fund the management and protection of these parks. Governments' actions have shown that they are unlikely to step in to fund any deficits because Africa's rapidly rising populations have changed the continental governments' spending priorities. Issues like health, education, housing, infrastructure and transport influence voters and are where budgets are being spent with little left to pay the costs to manage and conserve wild places. An example is South African National Parks, one of the best-funded conservation organisations on the continent. Yet SANParks receives only around 26% of its funding from the state. The 74% shortfall comes from revenues from tourism in various forms.[1] There are many other parks elsewhere in Africa that rely solely on tourism's revenues to pay for all their management costs.

It's not often that going on holiday makes a tourist one of the most effective conservationists. But that is mostly so when people travel to Africa today, a continent that receives only 7.4% of global tourism arrivals.[2] That's a remarkably low percentage, considering that Africa's total landmass is larger than all of the USA, China, India, much of Europe, and Japan combined. So what is holding Africa's tourism arrivals back despite the continent's unparalleled natural resources, size and one of the highest repeat and referral rates for any holiday destination anywhere? One of the handbrakes has been the recent 'flight shaming' trend, a movement that discourages people from taking flights because of their high carbon footprint and its impact on climate change. Yet Africa has a solid case for a flight-shaming exemption – fewer tourists to this continent could actually exacerbate the climate change crisis, as I will explain. Naturally, all flights cause harmful emissions; but, as flying globally contributes around 2.5% of global carbon emissions and, because Africa receives only 7.4% of global

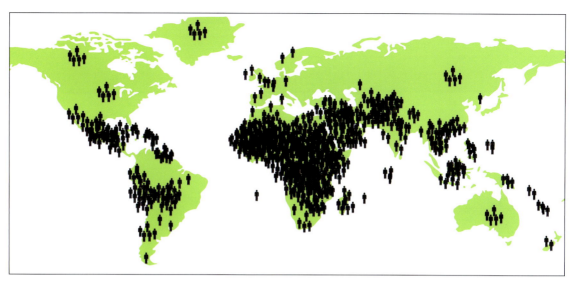

The icons highlight where the next billion people are expected to live, which will place increasing pressure on Africa's wildlife and wilderness areas. Wildlife authorities and the tourism sector must factor neighbouring communities into their plans and business models to ensure a sustainable future for wildlife. If this is not done, the very existence of wildlife could be at risk
Source: World Health Organization

tourists, flights to Africa constitute just a tiny portion of all global CO_2 emissions. Flight shamers could see the animal-rich wild savannas they wish to protect turned over to the plough in the absence of tourist revenues. That would be a climate disaster.

Africa's vast wild places – its forests, savannas and soils – are some of the most important carbon sequesters on the planet, storing much of the carbon produced by industrialised nations and providing vast amounts of oxygen in return. Contrary to popular belief, forests, while important, are not the most important sequesters of carbon – far more carbon is stored in savanna grasslands. But to process and store carbon at scale, these savannas have to remain pristine and natural. Any change in the landscape through farming, villages, people, development and even over-tourism rapidly diminishes their carbon-storage capacity. Africa's wild places need guaranteed long-term nurturing and protection so that they remain pristine for lions, for climate change and for the benefit of every person on the planet. The negatives from air travel to Africa are minuscule but the positive impacts from responsible tourism are immense.

But there is a flip side. Over-tourism is plaguing many of the world's most popular tourist destinations and is starting to rear its head in a few of the better known East and southern African wildlife destinations, where managing arrival numbers has to become a priority. There are a handful of parks and destinations that do receive way too many tourists, especially during the June-to-October peak season. The Maasai Mara, Amboseli, Ngorongoro Crater, Mana Pools, Savuti and Chobe all endure times of the year when there are way too many visitors – yet they are almost empty at other times. Plans currently underway to increase the number of permanent lodges in the Serengeti by 250% and mobile camps by 300% will certainly add the Serengeti to this list. The UNESCO World Heritage Site authorities and IUCN have noted that 'the increasing density of lodges, tented camps and other tourism infrastructure along the migration routes in the wider Serengeti ecosystem is increasingly likely to impact the wildebeest migration.' .This will worsen the detrimental impacts already caused by tourism infrastructure. To stop over-tourism, the authorities right around Africa need to use the best available technology to determine carrying capacity and put caps on lodges, hotels, mobile camps and visitors, especially over peak times. Tour operators need to do their bit by channelling customers away from overcrowded parks during the busy months or, better still, sending them to the way-off parks and reserves in need of tourism revenues.

Ngorongoro Crater, Nogorongoro Conservation Area, Tanzania.
© Martin Henfield

The low-volume, low-impact, high-revenue, high-job creation tourism model pioneered by Botswana and now being copied elsewhere in Africa is proving to be the most effective way to create the most benefits for the country and surrounding communities with the least impact on the environment.

Singita Grumeti, Tanzania. © Ross Couper

The last word:

Rewilding, the planet's last great hope

To conserve biological diversity is an investment in immortality. – EO Wilson

Colin Bell & Don Pinnock

When we were born, around 450,000 lions roamed the African plains, free and wild. Today, there are no more than 25,000 wild lions. This is a 95% drop in numbers … just in our lifetime. What if this plummet continues? What will be left for our grandchildren to experience in the upcoming decades? That concern is what led to the compiling and publishing of this book.

While the total extirpation of wild lions is too ghastly to contemplate, there is real hope – rewilding! This is the process that restores land to its natural uncultivated and pristine state. Despite all the bad news, wilding has been a common thread of hope throughout this book and, indeed, throughout the natural world where restoration of wildernesses has occurred with great success.

Many places on Earth have proven that when tracts of land are left to nature and indigenous wild species, the environmental clock is perfectly capable of being quickly reset to repair the consequences of poor land practices. If a small 450ha parcel of heavily cultivated land in Sussex in the UK is able to be successfully rewilded, with many species thought to be extinct or highly endangered returning, there is real hope for Africa and its lions.[1]

Where there is the will to return intensely farmed land and areas under human habitation to nature (and lions), it can be done. Majete National Park (see chapter 27) in Malawi, along with many of the great private game reserves in South Africa like Kwandwe, Kariega, Shamwari, Amakhala, Welgevonden, Marataba, Lapalala, Pilanesberg, Madikwe, Babanango, Phinda and the Makuleke/Pafuri area in northern Kruger National Park are all examples of this transformation. Majete has to be the shining star of the rewilding world, where lions and all the Big Five are once again roaming free and wild.

Rewilding in Africa is not only good news for lions, but also for climate change and the health of the planet. The messages to the world's leaders from scientists are clear, climate change is real, changes need to be made now and the challenge is immense.

Africa is vast and its wild places, its forests, savannas and their soils are some of the most important carbon sequesters on Earth. Africa's wild areas remove carbon from the atmosphere while producing a huge amount of oxygen for every species, hence the need for their guaranteed long-term protection. Africa has these soils in abundance in the continent's parks, reserves and wilderness areas where carbon emissions are accumulated, processed and neutralised for the good of the planet.

But preserving, managing, expanding and creating parks and reserves cost money. With the pressure of increasing numbers of people in need of basic services, governments throughout Africa can ill afford to prioritise the management and conservation of the continent's wildlife parks and reserves. Instead, those who vote them into power have to take precedence.

The funds needed for conservation are now derived, not from governments, but largely from tourism revenues, park entrance fees, concession fees and private grants. To illustrate this point, South African National Parks is one of the best funded and managed conservation organisations on the continent. Yet, it receives only 24% of its funding from the state.

Going on safari makes a traveller an extremely effective conservationist when they travel responsibly. We can argue flights to Africa do cause emissions. Yes, they do, but flying globally contributes around 2.5% of global carbon emissions. And, because Africa receives only 7.5% of global tourism arrivals, the negatives from tourist air travel to Africa are minuscule, while the positive impacts from safari tourism are immense.

The COVID-19 era proved Africa's parks and reserves need more than mere tourism revenues to fund the costs of managing and maintaining them. There needs to be a more reliable source of funding. This is where carbon and biodiversity credits should play their part; but, sadly, they don't. It's time for a global shakeup on this front.

So, how does the planet solve the dual crises of climate change and mass wildlife extinctions that threaten to forever change our world? In December 2022, over 190 countries adopted the Kunming-Montreal Global Biodiversity Framework (known as the Biodiversity Plan). For the first time in human history, the world has come together with a common goal to protect nature. This is an international commitment to better protect the planet that sustains us all.

One of the 23 targets aimed at reversing habitat and species loss (informally known as 30x30) specifically calls for the effective protection and management of 30% of the world's terrestrial, inland water, coastal and marine areas by the year 2030.

© Steph Vermeulen EQSA

This 30x30 is the biggest conservation commitment the world has ever seen to date and is effectively a rewilding promise.

If the politicians and bureaucrats live up to their word so we do effectively protect 30% of the planet's oceans, lands and freshwaters by 2030, people and nature will be able to survive and thrive together well into the future. Yet, for this to occur, rewilding is going to have to happen at a massive scale, and quickly. And if it happens in Africa, there will be real hope that lions and their ranges and numbers could well expand again. 30x30 could potentially be the roadmap to a nature-positive and lion-friendly future.

Action is needed and the converse is too dreadful to contemplate. As Wangari Maathai, the famous Kenyan environmentalist and Nobel Prize winner, said, 'The generation that destroys the environment is not the generation that pays the price. That is the problem.'

Serengeti National Park, Tanzania. © Dana Allen

Serengeti National Park, Tanzania. © Dana Allen

Kruger National Park, South Africa. © Marlon du Toit

Serengeti National Park, Tanzania.
© Johan van Zyl

A plea to the next generation

Virginia McKenna OBE
Born Free Foundation

I have always admired, respected and valued animals – of whatever species. I must admit, I have had special feelings for lions for an exceptionally long time, perhaps because I had the privilege to meet and get to know several when my husband, Bill Travers, and I worked on the film *Born Free* in 1964.

In those far-off *Born Free* days, Bill and I got to know five lions very well; walking on the plains in the early morning light, resting under a tree, swimming in the ocean. These are memories like no other. The care the females give their young, the bonding of family members, their loyalty to the pride.

The deeply shocking realisation that, if we are not vigilant, they could be lost from across much of their African range is unthinkable. From perhaps 200,000 when we made the film, to around 20,000 today is a statistic we must all take to heart and act upon. Lions could be extinct across vast swathes of the continent by 2050. To say I have always been appalled by the practice of trophy hunting is an understatement. That anyone could have pleasure or pride in killing a beautiful living creature is beyond understanding. A head on a wall or a rug on a floor is surely something to be deeply ashamed of.

Whenever I have been fortunate enough to see wild lions on safari, I have experienced some of the most wonderful moments in my life. I realise, of course, not everyone can do this but, today, amazing filmmakers can bring stories about the life of lions into our homes. Those images should inspire us. Yet, alone, they are not enough – we must do more. Being an optimist, I always have hope. And now, in 2024, I sense there is a growing number of young people who have deep feelings for all creatures that have no voice.

'I am the voice of the voiceless. Through me the dumb shall speak.' Those words by Ella Wheeler-Wilcox say it all and we must remember and act on them always.

About the authors

We asked scientists, game guards, activists, academics, journalists, lodge owners and leaders of NGOs involved in lion work throughout Africa if they would be prepared to write chapters for this book. The response was extraordinary and heartening. We are indebted to all these people who took time to pen the thousands of words that make up this book. They have done it free of charge and for the creatures they love and respect: lions. We salute you and thank you.

Peter Anderson is a wildlife property specialist with over 25 years of experience in creating and enlarging conservation and protected areas across southern Africa. This included the establishment of sustainable projects around tourism and communities throughout the region. He has worked on the rewilding of several private protected areas with a current focus on Lapalala Wilderness and the Lapalala Wilderness School in the Waterberg Biosphere.

Chris Badger began his travel career as a driver/guide for Exodus Expeditions, leading trips through Nepal, India and Pakistan before transitioning to conducting trans-Africa trips from Tunisia to South Africa. In 1982, he joined Londolozi Game Reserve as a ranger, later expanding his tours to Namibia, South Africa, Zimbabwe and Botswana as a freelance guide. He subsequently joined Wilderness Safaris in 1986 and relocated to Malawi in 1987, where he has remained since. He and his wife Pam have three children, all raised in Malawi, who are now actively involved in conservation. (He is also a passionate blues musician.)

Drew Bantlin holds an MSc from the University of Wisconsin-Madison, where he researched the movements of reintroduced lions in Akagera National Park, Rwanda. He is currently the Regional Conservation Manager for Rwanda and Malawi at African Parks.

Prof Hans Bauer is a scientist with three decades of experience in lion conservation, particularly across West, Central and the Horn of Africa. He has written many landmark publications, including the IUCN Red List Assessment for lions. Working for the University of Oxford but residing in the region, he has a focus on human-lion conflict, local capacity strengthening and surveys of remote wild places.
www.wildcru.org

Dr Colleen Begg is a South African conservation ecologist and photographer with a PhD in zoology and more than 30 years' experience in human-wildlife conflict and carnivore conservation. She is also the Managing Director of the Niassa Carnivore Project, on the leadership team of the Wildlife Conservation Network's Lion Recovery Fund and a co-founder of Women for the Environment, Africa. She has spent the past 20 years in the field in Niassa Reserve, Mozambique, and is particularly interested in finding locally derived conservation solutions that are inclusive and sustainable.
www.niassalion.org and www.womenforenvironment.org

Keith Begg is the Operations Director of the Niassa Carnivore Project, Director of TRT Conservation Foundation, a pilot and an experienced filmmaker and photographer. He is a South African conservationist with extensive experience in resolving conflicts between people and carnivores, from honey badgers to lions. He has spent the past 20 years in Niassa Reserve, northern Mozambique, documenting and implementing conservation solutions with local communities.
www.niassalion.org

You could say **Colin Bell** made his own luck – instead of going into the world of finance after completing his economics degree at Wits University, he landed his first job as a safari guide in Botswana in 1977. In 1983 he co-founded Wilderness Safaris and later co-founded both Great Plains Conservation and Natural Selection Safaris in 2006 and 2013 respectively. Colin has co-authored two books on wildlife, the environment and sustainable tourism – *Africa's Finest* and *The Last Elephants*. The companies that he founded run camps and lodges across nine African countries and successfully and sustainably manage millions of hectares of prime wildlife reserves.

Laura Bertola is a scientist and conservationist working in the field of conservation genetics with a focus primarily on carnivores and other African mammals. She is interested in how we can translate genomic resources into insights and tools that have relevance and applicability for the conservation of biodiversity.
https://laurabertola1.wixsite.com/mysite

Natalia Borrego is a postdoctoral researcher at the Max Planck Institute of Animal Behaviour and the Univeristy of Minnesota's Lion Center. She is also project leader at the University of Konstanz Centre for the Advanced Study of Collective Behaviour. She has been studying African lions for over a decade and is a member of the Lion Management Forum in South Africa. Her research focuses on lion cognition, sociality and behavioral ecology.
www.saiia.org.za

André Botha is the manager of the Endangered Wildlife Trust's Vultures for Africa Programme and has worked in conservation for more than 30 years. He also co-chairs the IUCN Species Survival Commission's Vulture Specialist Group and focuses on combatting the impact of wildlife poisoning and the implementation of the multi-species action plan for African-Eurasian Vultures across three continents.

James Byrne worked at National Geographic Television from 2000 to 2012, producing and writing TV programmes about wildlife, conservation, science and culture. In 2009, he made Nat Geo's first film about Gorongosa, *Africa's Lost Eden*, and fell in love with the place and the project. In 2012, he joined the Greg C. Carr Foundation/Gorongosa Project as Media Director, where he leads a small in-house team that documents the ongoing story of Gorongosa's conservation and human development work.

Emma Childs is an owner of The Emakoko in Nairobi National Park. She has lived and worked with wildlife for 26 years and resides in the national park with her family. The park lions have been a part of her and her family's lives for 12 years. She has seen the rise and fall of two of the park's great lion kings and has had, she says, far too many encounters with some of the park's lionesses.

Chad Cocking has been a professional field guide and wildlife photographer based in the Timbavati Private Nature Reserve in South Africa's Greater Kruger region since 2007. He developed a deep fascination with the region's big cats and has spent close to two decades following many of the established lion prides from one generation to the next. Within these prides, he has been privileged enough to see 10 different naturally-occurring white lions, carriers of the rare recessive genetic mutation that put the Timbavati on the map in the 1970s.

Sally Capper is CEO of Carbon Tanzania, an impact-driven social enterprise that uses forest conservation to combat climate change. From 2020-2024 she served as Director of Development & Strategy of KopeLion, driving community-led conservation efforts to enable long-lasting coexistence between people and lions in Ngorongoro, Tanzania.

Dr Samuel Cushman is a Senior Fellow in the Wildlife Conservation Research Unit at the University of Oxford. He studies a range of topics in spatial ecology and conservation, including landscape pattern analysis, landscape dynamic simulation modelling, landscape genetics, movement and connectivity ecology, species distribution and habitat relationships modelling, and scenario optimisation for natural resources management and conservation.

Chris Fallows is a multi-award-winning photographer whose work over the past three decades has focused on the planet's most iconic species. **Monique Fallows** is a naturalist who has worked in the field with great white sharks and other megafauna, both on the land and sea, for the past 25 years. She is also a children's book author.

Chiara Fraticelli is a conservationist who worked for several years with the African Parks Network in the Greater Zakouma Ecosystem in Chad. As a PhD student, she is researching lion ecology, seasonal movement and human-lion conflict in southern Chad.

Laurence Frank is an internationally regarded large carnivore biologist with a lifetime focus on lions and savanna ecosystems. His PhD was on predator-prey relationships and territorial behaviour of African lions in the Kruger National Park. Thereafter he spent several years as an independent consultant biologist, managing multi-disciplinary research programmes contributing to the management of large national parks in southern Africa. He is an associate professor at Tshwane University of Technology and from 2013 to 2022 he was Senior Director of Panthera's Lion Programme, based in the vast Kavango-Zambezi Transfrontier Conservation area, working mostly in Namibia, Angola and in Kafue and Hwange national parks. Paul has published over 70 scientific papers, two books, many reports and has presented at many international conferences. He has also scripted and been interviewed for several documentary movies.

Dr Paul Funston is a senior research scholar in Rangeland Ecology at the Okavango Research Institute in Maun, Botswana. His research is on grazing ecosystem ecology and he holds a PhD from the University of KwaZulu-Natal, South Africa.
www.ori.ub.bw

Chris Galliers is president of the International Ranger Federation. Based in South Africa, he has a passion for conservation and the people tasked to care for it. This has led him to work in the protected area management and development space, working for Conservation Outcomes, as well as with rangers across the continent as Chairperson of the Game Rangers Association of Africa.

Jeremy Goss was born in South Africa, and the continent's wild spaces have always been part of his life. This fuelled an understanding that few such spaces are safe and with that came a growing desire to contribute to conservation. After completing an MSc in Conservation Biology, he moved to Kenya to research human-predator conflict in the Amboseli ecosystem, where he has been living and working for Big Life Foundation since 2013.

Ryan Hillier was born and raised in Zimbabwe. He followed his dream of becoming a guide, and began working at Kwandwe Private Game Reserve in 2009, where he is still based. Here, he met his wife, and they now have two children, who they are privileged to raise out in the bush. He has always been fascinated by big cats, especially lions and the dynamics between individuals and rival prides.

Dr Neil R Jordan is a Senior Lecturer at UNSW Sydney in a joint position with Taronga Conservation Society Australia, and collaborates with Botswana Predator Conservation.

Agostinho Jorge is a Mozambican conservation ecologist specialising in large carnivore conservation at the interface between people and carnivores. He completed his MSc on the costs and benefits to local communities of the presence of leopards and is currently completing his PhD with KwaZulu-Natal University on the bushmeat trade in Niassa Reserve. He is the Conservation Director of the Niassa Carnivore Project and co-leads a team of more than 100 Mozambicans involved in carnivore and community conservation in Niassa Reserve, Mozambique. He has a particular interest in inclusive, participatory leadership in conservation and is a member of the Leadership Alumni Committee of Maliasili. www.niassalion.org

Dr Żaneta Kaszta is an Assistant Research Professor at Northern Arizona University's School of Informatics, Computing, and Cyber Systems. Her research focuses on landscape ecology and genetics, species distribution and biodiversity modelling, decision support systems for conservation, the impact of land use change on wildlife and ecosystems, tropical ecology and remote sensing.

Pieter Kat has a PhD in ecology and evolution and has worked on conservation issues for a diversity of African species, among them the genetics of antelopes, rabies viruses, feline immunodeficiency virus among lions and the ecology of African wild dogs, jackals and lions, as well as the evolution of invertebrate species in the African great lakes and established international conservation programmes with funding from the UK, EU and USA. He is a founder and trustee of the UK-based lion conservation charity LionAid and actively lobbies for better lion conservation in a range of African nations, including Zambia, Kenya and Tanzania, as well as implementing projects aimed at reducing lion-livestock conflict.

Saning'o Kimani is the Programs Coordinator at Kopelion, where he leads initiatives focused on human–lion coexistence. Born and raised in the Ngorongoro Conservation Area, his deep connection to the land, community and wildlife shapes his efforts to create a sustainable balance between people and other species in the region.

Jonathan L Kwiyega is a co-founder and Executive Director of the grassroots non-profit Landscape and Conservation Mentors Organization in western Tanzania. He has an MSc in Natural Resource Management, is a passionate wildlife conservationist and local educator promoting human-lion coexistence.

Graeme Lemon was brought up in the Zambezi Valley in northwestern Zimbabwe. He is a professional guide, conservationist and adventurer. He is based in Ethiopia from where he runs 'off the beaten track' safaris to fascinating destinations in the Horn of Africa.

Andrew Loveridge is Director of the Lion Programme at Panthera and an Associate Professor at Oxford University. Much of his time over the last 25 years has been spent running an extensive, long-term lion research and conservation programme spanning 17 protected areas in three international transboundary areas critical to lion conservation, A successful scientist and author, he has published over 150 peer-reviewed articles. He is a member of the IUCN SSC Cat Specialist Group and IUCN African Lion Working Group and a contributor to IUCN SSC *Panthera leo* species conservation guidelines and IUCN Red List assessment.

Prof David Macdonald has a background in behavioural ecology, with an emphasis on carnivores, although his research has spanned published studies on organisms from moths to penguins and plants. He is Oxford University's first Professor of Wildlife Conservation and founded the Wildlife Conservation Research Unit (WildCRU). He was ranked as number 26 of 10,000 most influential researchers in the disciplines of Ecology and Evolution. He has received many awards and was elected a Fellow of the Royal Society of Edinburgh, and a Conservation Fellow of the Zoological Society of London. In 2010, he was honoured with the title of Commander of the Order of the British Empire (CBE).

Ingrid (Inki) Mandt was born and bred on a cattle farm in Otjiwarongo in Namibia. A retired dentist, she shares her love for wildlife and photography with her husband Izak Smit, spending much time in the wild. She developed a fascination with the desert-adapted lions in Damaraland and Kaokoland. She has spent many hours getting to know, photograph and protect several generations of desert lions and has developed a deep insight into their precarious existence and the threats they face.

Ian Michler has spent over 33 years working across Africa as a tourism operator, environmental photojournalist, wilderness guide and ecotourism consultant. He is a graduate of the Sustainability Institute, Stellenbosch University and is a director of *Blood Lions* documentary and the global campaign to end predator breeding and canned hunting.
www.inventafrica.com

Belinda J Mligo has over three years of field experience in the wildlife and public health field. She is the Human-Lion Coexistence Programme Coordinator at the WASIMA People, Lion, Environment NGO. She focuses on community conservation education and outreach campaigns and projects for human-lion coexistence among communities living near protected areas in western Tanzania. She has worked as a research assistant in various wildlife and One Health projects and has volunteered at National Research Medical Institute as an enumerator. Belinda holds an MA in Public Health and Food Safety and a Bachelor's degree in Wildlife Management.

Tutilo Mudumba is a researcher and lecturer at Makerere University, specialising in fields such as carnivore ecology, biodiversity conservation, human dimensions of wildlife and GIS applications. Tutilo holds executive roles in various organisations and is an Expert – Research and Monitoring at Uganda Wildlife Authority.

Olivier Mukisya serves as the spokesperson for Virunga National Park, a UNESCO World Heritage Site and a sanctuary for endangered mountain gorillas.

Thandiwe Mweetwa is a wildlife conservationist working in eastern Zambia. Her areas of interest include carnivore research, large landscape conservation, community engagement, human-wildlife conflict mitigation, diversity inclusion in conservation and youth-leadership development.

Bhushan Pandya is a wildlife photographer in India and a former member of the State Board for Wildlife in Gujarat. He has travelled to about 30 protected areas in Africa, Nepal and India and has documented 18 wildlife censuses plus the official documentation of Gir National Park, Gir Wildlife Sanctuary and Girnar Wildlife Sanctuary. His photographs and articles have been widely published in books, newspapers, international journals and magazines and he has worked with documentary teams filming *Animal Planet*, BBC/Discovery and National Geographic TV channels.

Don Pinnock is a historian, criminologist and environmental journalist who had the good fortune to work, first as a writer/photographer, then as editor of the travel magazine *Getaway*. For more than a decade he explored much of Africa with his camera and pen, meeting extraordinary people and writing what turned out to be prize-winning features. Along the way he fell in love with the natural world. Don is the author of 17 books. He has won two Mondi Awards for his environmental columns, the City Press Non-Fiction Award for his book *Gang Town*, and his novel on Khoisan magic, *Rainmaker*, was shortlisted for the European Union Literary Award.

David Quammen is an author and journalist whose work has appeared in *National Geographic*, *The New Yorker* and other magazines. His most recent books are *Breathless* (about the COVID virus) and *The Heartbeat of the Wild* (about wild creatures, wild places and their conservation). He received widespread acclaim for his groundbreaking book on biogeography, *Song of the Dodo*.

Andrea Reid began her career in Hluhluwe iMfolozi Game Reserve, South Africa, gaining a variety of experience in a number of programmes and roles. In 2007, she joined African Parks in Liuwa National Park, Zambia, assisting with tourism and park operations. Since 2012, she has worked as Special Projects Manager, working in Bangweulu Wetlands, Zambia, Liwonde National Park, Malawi and Kafue National Park, Zambia.

Craig Reid holds a National Diploma in Nature Conservation and has over 30 years' experience in Protected Area Management. He worked in iMfolozi Game Reserve, South Africa, and joined African Parks in 2007. Craig has managed Liuwa Plain National Park and Bangweulu Wetlands in Zambia and led the initiation of the African Parks project in Liwonde National Park, Malawi, and Kafue National Park, Zambia.

Rob Reid is a freelance consultant who served as Park Manager for the African Parks Network from 2014 to 2017, and as Field Operations Manager from 2017 to 2019.

Super Sande has literally been living in the Botswana bush his whole life. Born in Senyawe, a village in North-East Botswana, he started working for Jack Bousfield as a teenager at an animal orphanage in Francistown. After Jack passed away in 1992, Super started working for Ralph Bousfield and helped him build Jack's Camp. Ralph saw Super's potential and introduced him to the world of guiding and over the years Super's diligence and passion has made him a true specialist in this challenging and unique environment and ecosystem, culminating in being awarded the honour of being named the best guide in Africa for 2023. Super has incredibly good tracking skills and eyesight like a meerkat, which combined make him an expert on finding the Kalahari's most elusive residents. He truly loves the Makgadikgadi Pans and the challenges that come with guiding there. There is nowhere on earth quite like this ancient super-lake and no one on earth who knows it better to write about it than Super.

Jonathan & Angela Scott are award-winning wildlife photographers, television presenters and authors of 40 books. In 2021 they founded the Sacred Nature Initiative (SNI) to help reconnect people to nature. Their large portfolio books include *Sacred Nature: Life's Eternal Dance* and *Sacred Nature Volume 2: Reconnecting People to Our Planet*. Their most recent television series *Big Cat Tales* (Seasons 1 and 2) are on Animal Planet. The documentary *Lion: The Rise and Fall of the Marsh Pride* charts the history of the pride that Jonathan first started to follow in 1977. They are the only couple to have won the Overall Award in the prestigious Wildlife Photographer of the Year Competition as individuals – Jonathan in 1987 and Angela in 2002.
www.jonthanangelascott.com and www.sacrednatureinitiative.com

Izak Smit is originally from Cape Town and now living in Swakopmund. He is a businessman and retired commercial pilot flying helicopters, gyroplanes and fixed-wing aircraft in Africa. He and his wife Inki spent three decades getting to know Damaraland while pursuing their love for photography and desert wildlife. Time spent in the home ranges of the desert-adapted lions resulted in their being approached by Conservancy Communities 12 years ago to assist with human-lion conflict mitigation, which in turn led to their establishing their NGO, Desert Lion Human Relations Aid.

Dr Lara Sousa works at the University of Oxford's Wildlife Conservation Research Unit (WildCRU). She is a member of the Trans-Kalahari Predator Programme research and is investigating the spatial ecology of different species of carnivores in Zimbabwe, focusing on the estimation of population densities of three species (lion/leopard/hyaena). Her research interests include the behavioural and spatial ecology and species interactions of wild vertebrates.

Brent Stapelkamp is a Zimbabwean who found that his calling was to work for the good of lions and has devoted most of his adult life to carrying out this mission, as well as working with the people who live with such wildlife. Co-founder of The Soft Foot Alliance Trust, he works with his wife Laurie in creating landscapes of abundance to mitigate conflict with wildlife.

Brent Stirton is a South African photographer with an extensive history in the documentary world. He is a fellow of the National Geographic society and a National Geographic explorer. Brent has received awards from World Press Photo Foundation and the Pictures of the Year International competition, *National Geographic* magazine's Photographer's Photographer, the Overseas Press Club, the Webbys, the Association for International Broadcasting, the HIPA Awards, the Frontline Club, the Deadline Club, *DAYS JAPAN*, China International Press Photo Contest, Germany's Lead Awards, Graphis, *Communication Arts, American Photography, American Photo* and the American Society of Publication Designers as well as the London-based Association of Photographers. He won a National Magazine Award for his work for *National Geographic* magazine in the Congo and has been recognised by the United Nations for his work on the environment and in the field of HIV/AIDS.

Lindy Sutherland is CEO of the Kariega Foundation and family owner of Kariega Game Reserve. She is a passionate conservationist and is driven to achieve sustainable protection of biodiversity through community collaboration and development.

Marcus Westberg is an internationally renowned, award-winning conservation photojournalist, and a Senior Fellow of the International League of Conservation Photographers. His work focuses primarily on the complex yet crucially important relationship(s) between humans and the natural world in sub-Saharan Africa.

Jeffrey Wu is a Canadian professional wildlife and nature photographer, author, educator and conservationist. His love of photography was nurtured by his mother, a professional photographer who taught him photography from the age of seven. Jeffrey emmigrated from Shanghai, China to Canada, and in 2012 he went to Kenya on a photography trip that would change his life forever. In 2013 he sold his restaurants in Toronto to become a full-time professional wildlife photographer, spending 10 years mainly photographing wildlife in the Maasai Mara. He has judged some of the most prestigious competitions of the world, including the Nikon Photo Contest and Nature's Best Photography Africa. He is the Kenya Tourism Board Brand Partner and his works and articles have been published in more than 120 publications worldwide including *Africa Geographic*, *Chinese National Geography*, *SWARA*, *People*, *Canadian Camera*, *Remembering Cheetahs*, *Outdoor Photography Canada*, *The Times*, *Daily Telegraph*, *Daily Mail*, *Cultural Geography* and *The Daily Star*.

Acknowledgements

This book is the work of many people who gave their time and knowledge and shared their passion for lions and Africa's fragile environments. These include 34 specialists from across the continent and 86 wildlife photographers and safari guides, without whom this book would not have been possible. They are all exceptional conservationists and artists in their field. Special thanks goes to Dr Paul Funston, whose extensive knowledge in the field of lion research at the start of this project helped identify who to approach to contribute chapters. Special thanks, also, to internationally acclaimed photographers Brent Stirton and Jeffrey Wu for their extraordinary portfolios and to internationally acclaimed conservation writer David Quammen for agreeing to write the Foreword.

Lions are under severe threat from human depredation across the planet and the people and organisations in this book are heroes holding the line against their extinction.

We have also been fortunate to have patient and highly skilled partners in the construction of this book. They include designers Verena Altern and Neil Bester, copy editors Margy Beves-Gibson and Helen de Villiers, and the team at Struik Nature – Colette Alves, Roelien Theron and our publisher, Pippa Parker – whose patience and forbearance on the long road to the final book was way beyond the call of duty.

We would also like to thank Smithsonian Books as international partners in the book's production and distribution.

And lastly, we thank those extraordinary, majestical creatures – lions – which have inspired and wowed us at every step of the way. May they live long, stay safe and multiply. A world without them would be unthinkable.

References

Chapter 1: Where have all the lions gone?
Dr Andrew Loveridge, Dr Lara Sousa, Dr Samuel Cushman, Dr Żaneta Kaszta & Prof David Macdonald

1: Ceballos & Ehrlich (2002); Ripple et al (2014).
2: Guggisberg (1961).
3: Tilman et al (2017).
4: Pauly (1995); Plumeridge & Roberts (2017).
5: Bauer et al (2015); Morrison et al (2007).
6: Henschel et al (2015).
7: Curry et al (2020); Dures et al (2019).
8: Hazzah et al (2009); Main (2020).
9: IUCN-SSC (2006); Panthera (2021).
10: United Nations (2019).
11: Malbrant & Maclatchy (1949), in Chardonnet (2002).
12: Hedwig et al (2018); Henschel et al (2014).
13: Barnett et al (2018).
14: Crooks et al (2017).
15: Reed & Frankham (2003).
16: De Manuel et al (2020); Van de Kerk et al (2019).
17: Onorato et al (2010); Packer et al (1991); Wildt et al (1987).
18: Trinkel et al (2011).
19: Antunes et al (2008); Curry et al (2020); Dures et al (2019).
20: Curry et al (2020).
21: Curry et al (2019); Dubach et al (2013); Morandin et al (2014); Smitz et al (2018); Tende et al (2014).
22: Bongaarts (2009); Holdren & Ehrlich (1974).
23: United Nations (2019).
24: Díaz et al (2019); Tilman et al (2017).
25: Laurance et al (2014).
26: Bradshaw & Di Minin (2019).
27: Newmark (2008); Östberg et al (2018).
28: Di Minin et al (2021).
29: Lindsey et al (2018); Robson et al (2021).
30: Bertola et al (2021).
31: Miller et al (2015).
32: Wittemyer et al (2008).
33: Loveridge et al (2017).
34: Elliot et al (2014); Matshisela et al (2021).
35: Estrada-Carmona et al (2014); Kaszta et al (2020a).
36: Marchini et al (2019); Sibanda et al (2021).
37: Schroth & McNeely (2011).
38: Kaszta et al (2020b).

Chapter 2: The most charismatic of cats
Dr Natalia Borrego

1: Packer et al (2001).
2: Borrego & Gaines (2016).
3: Morapedi et al (2021).
4: O'Connor et al (2022).
5: Blackwell et al (2016).
6: Miller & Funston (2014).

Chapter 3: Saving Africa's lions
Dr Pieter Kat

1: Curry et al (2020). https://academic.oup.com/mbe/article/38/1/48/5871931.
2: Adams & Carwardine (2009).

Chapter 4: Genetic variability in lions
Dr Laura Bertola

1: Ceballos et al (2017).
2: Burger et al (2004).
3: Stuart & Lister (2011).
4: Yamaguchi et al (2004).
5: Schnitzler (2011).
6: Riggio et al (2013).
7: Loveridge et al (2022).
8: Barnett et al (2006); Bertola et al (2011).
9: Loveridge et al (2022).
10: Barnett et al (2014); Bertola et al (2015, 2016, 2022b).
11: Kitchener et al (2017).
12: Robson et al (2021).
13: Bertola et al (2015, 2022b).
14: Barnett et al (2006); Bertola et al (2011).
15: De Vivo & Carmignotto (2004); Bertola et al (2016); Kingdon (2018).
16: Bertola et al (2016).
17: Bertola et al, in prep.
18: Trinkel et al (2008).
19: Bertola et al (2021).
20: Becker et al (2022).
21: Bertola et al (2021).
22: Bertola et al (2022a).
23: Spong et al, in prep.
24: Wasser et al (2010).
25: Ceballos et al (2017).

Chapter 5: The poison problem
André Botha

1: https://pubmed.ncbi.nlm.nih.gov/24716788/.
2: https://illegalwildlifetrade.net/2019/04/08/traditional-medicine-and-iwt-in-southern-africa-a-roaring-trade.
3: https://abcnews.go.com/International/south-africa-end-captive-lion-breeding-bone-trade/story?id=77451913.
4: https://www.citizen.co.za/ridge-times/lnn/article/elephant-46-vultures-and-4-lions-killed-in-knp/.
5: https://bloodlions.org/demand-lions-bones-poisonous-affair/.
6: https://niassalion.org/threats/.
7: nationalgeographic.com/animals/article/wildlife-watch-lions-poisoned-uganda-cattle-retaliation.
8: www.africanwildlifepoisoning.org/.
9: www.africanparks.org/african-wild-dogs-poisoned-liwonde-national-park-malawi.
10: www.wcs.org/wildcards/posts/a-poisoning-in-ruaha.
11: www.pressreader.com/zimbabwe/chronicle-zimbabwe/20160204/281663959044395.
12: https://africageographic.com/stories/mass-poisoning-leaves-lions-vultures-dead-ruaha/.
13: https://ewt.org.za/what-we-do/saving-species/vultures/.
14: https://peregrinefund.org/africa-program-coexistence-co-op.
15: www.aa.com.tr/en/africa/uganda-jails-2-poachers-for-17-years-over-killing-of-lions/2674884.
16: www.zambiacarnivores.org/conservation-action.

Chapter 6: Lions of the north
Prof Hans Bauer

1: Bauer et al (2018).
2: Bertola, et al (2019).
3: Bauer et al (2015).
4: www.newscientist.com/article/mg17223150-700-lions-in-peril/.
5: www.science.org/content/article/lions-trouble.
6: Henschel et al (2014).
7: Lhoest et al (2022).
8: Bauer & De Iongh (2005).
9: Bauer et al (2010).
10: Aebischer et al (2020).
11: Mohammed et al (2023).
12: Bauer et al (2021).

Chapter 7: Building a lion stronghold
Chiara Fraticelli

1: Vanherle (2011).
2: Olléová & Dogringar (2013).

Chapter 11: A newly rediscovered migration
Marcus Westberg & African Parks

1: Schapira et al (2016)

Chapter 12: Maasai Mara - a fragile Eden
Jonathan & Angela Scott

1: Milton (1997).
2: Vettorazzi et al (2022).

Chapter 13: Lions return to Amboseli
Jeremy Goss

1: Hazzah et al (2009).

Chapter 14: Lions on the doorstep
Emma Childs

1: Chicago Tribune (1996). www.chicagotribune.com/news/ct-xpm-1996-08-04-9608040033-story.html.
2: New World Encyclopedia, www.newworldencyclopedia.org/entry/Nairobi_National_Park.
3: Tsavo Trust (2021). https://tsavotrust.org/kenyas-first-ivory-burn-an-influential-moment-in-the-history-of-elephant-conservation/.

Chapter 21: Traditional use of lion parts
Jonathan L Kwiyega & Belinda J Mligo

1: Borgerhoff Mulder et al (2009).
2: Mole & Newton (2020).
3: Maclennan et al (2009); Ogada et al (2003); Kiffner et al (2009).
4: Mesochina et al (2010); Brink et al (2016).

5: Packer et al (2011); Riggio et al (2013).
6: Kwiyega et al (2020).
7: Kiffner et al (2009); Mesochina et al (2010); Fitzherbert et al (2014).
8: Kwiyega et al (2020).
9: Fitzherbert et al (2014).
10: Dickman (2008); Frank et al (2006); Holmern et al (2007); Kissui (2008), Patterson (2004).
11: Frank et al (2006); Packer et al (2005, 2006).
12: Whitman et al (2007).
13: Spear & Waller (1993); Wilson (1953); Kissui (2008).
14: Goldman et al (2010, 2013).
15: Borgerhoff Mulder et al (2009, 2019).
16: Loveridge et al (2010); Goodrich et al (2015).
17: Williams et al (2017).
18: Mole & Newton (2020).
19: Born Free Foundation (2008).

Chapter 22: Dam and be damned: Stiegler's Gorge
Colin Bell

1: www.juliusnyerere.org/about/mwalimu_nyerere_on_wildlife_conservation.
2: https://whc.unesco.org/en/list/199/.
3: Bauer et al (2016).
4: Personal communication.
5: www.defenceweb.co.za/security/border-security/tanzania-to-shut-part-of-wildlife-preserve-to-big-game-hunters/.
6: Pinnock & Bell (2019).
7: www.gga.org/tanzanias-very-expensive-white-elephant/.
8: www.bbc.com/news/world-africa-52966016.
9: www.iucn.org/news/iucn-42whc/201806/tanzania-urged-halt-logging-plans-and-dam-project-selous-game-reserve-advised-iucn.
10: https://allafrica.com/stories/201107170147.html.
11: https://e360.yale.edu/features/despite-warnings-a-destructive-african-dam-project-moves-ahead#.

Chapter 24: Where are Angola's lions?
Dr Paul Funston

1: IUCN (2006a).
2: Bauer & Van der Merwe (2004).
3: Bauer et al (2016).
4: IUCN (2006a).
5: Riggio et al (2013).

Chapter 30: Understanding coexistence
Dr Colleen Begg, Keith Begg & Dr Agostinho Jorge

1: Somerville (2020).
2: Dominguez & Luoma (2020).
3: Liesegang (2003).
4: Kushnir et al (2010).
5: Newitt (1995).
6: Israel (2009).

Chapter 32: Blood under the carpet
Brent Stapelkamp

1: Elliot (2014).
2: Loveridge et al (2017).

3: MacDonald et al (2015).
4: Loveridge et al, in prep.

Chapter 41: Bones of contention
Dr Don Pinnock

1: Yale Environment 360: 'The Ongoing Disgrace of South Africa's Captive-Bred Lion Trade'.
2: Ibid.
3: Personal communication.
4: The South African: 'South Africa breeds lions and tigers for traditional medicine'.
5: FOUR PAWS: The Vicious Cycle report.
6: Ibid.
7: Lion Aid: Roger Cook and *The Cook Report*.
8: Blood Lions: How the lid was lifted on canned lions.
9: Blood Lions: About the film.
10: Yale Environment 360, op cit.
11: EMS Foundation & Ban Animal Trading report: The Extinction Business.
12: Bloomberg Law: 'South Africa's High Court declares lion bone export quotas unlawful'.
13: Ibid.
14: Personal communication.
15: The Extinction Business, op cit.
16: OCCRP investigation: Inside South Africa's brutal lion bone trade.
17: The Vicious Cycle, op cit.
18: Ibid.
19: New York Times: 'In trafficking of wildlife, out of reach of the law'.
20: Quoted in Born Free Foundation report: Cash before Conservation.
21: Ibid.
22: The Extinction Business, op cit.
23: Ibid.
24: Paul O'Sullivan's website, Forensics for Justice.
25: OCCRP, op cit.
26: The Extinction Business, op cit.
27: Environmental Affairs Minister gazettes Biodiversity Management Plan for African lion for public comment.
28: https://pmg.org.za/files/190312Colloquium_resolutions.ppt.
29: The African Lion Conservation Community's Response to the South African Predator Association's Letter.
30: Hunter et al (2013); Bauer et al (2016).
31: SAIIA report: Picking a bone with captive predator breeding in South Africa; CACH UK & SPOTS Netherlands report: Captive lion breeding, canned lion hunting & the lion bone trade: Damaging Brand South Africa?
32: Environmental Investigation Agency report: The lion's share: South Africa's trade exacerbates demand for tiger parts and derivatives.
33: Humane Society International survey: South Africa National Public Opinion (2018).
34: DFFE: Publication for comments (2020).
35: Green et al (2020).
36: DFFE: High-Level Panel report: https://www.gov.za/sites/default/files/gcis_document/202106/44776gon566.pdf.
37: Consultation on the draft White Paper on conservation and sustainable use of South Africa's biodiversity 2022.
38: Heinrich et al (2022).

Chapter 42: Killing the king
Ian Michler

1: https://abcnews.go.com/International/south-africa-end-captive-lion-breeding-bone-trade/story?id=77451913#:~:text=%22The%20panel%20recommends%20that%20South,consultation%20in%20implementation%20is%20conducted.%22

Chapter 43: The last lions of Asia
Bhushan Pandya

1: A research paper was published in Cat News (2017), the newsletter of IUCN.
2: Divyabhanusinh (2014).
3: Bhanusinh(2006).
4: https://nja.gov.in/Concluded_Programmes/2016-17/SE-5_(25,26-03-2017)_PPTs/6.The%20Wildlife%20(Protection)%20Act,%201972.pdf.
5: Wildlife Division, Sasan-Gir report: Population Status of Wild Prey Species in Gir Protected Areas.
6: Ram et al (2023).
7: The Protected Areas and the lion home range in the multi-use landscape.
8: www.conservationindia.org/wp-content/files_mf/Lion-judgment-SC-Apr-2013.pdf.
9: Marshall, L 2006 The Last Lions of India. BBC.
10: Lion@2047: A Vision for Amrutkal.
11: www.hindustantimes.com/india-news/cyclone-tauktae-uprooted-over-3-5-million-trees-in-gir-101659681377898.html.
12: www.insightsonindia.com/2020/06/12/census-of-asiatic-lion/.
13: IUCN Red List of Threatened Species in 2008: www.iucnredlist.org/species/247279613/247284471.
14: IUCN Red List of Threatened Species in 2023: www.iucnredlist.org/species/15951/231696234.

Chapter 44: Under-tourism, flight shaming and the scourge of over-tourism
Colin Bell

1: South African National Parks Annual Performance Plan 2024/25: https://www.sanparks.org/wp-content/uploads/2024/06/SANParks-Annual-Performance-Plan-2024-2025.pdf
2: https://wttc.org/Portals/0/Documents/Reports/2023/Africa%20report/Africa-TandT-Growth-ExecSummary011123.pdf?ver=T8JgcY7wChgbOIRq0HFdGA%3D%3D

The last word: Rewilding, the planet's last great hope
Colin Bell & Don Pinnock

1: Tree, I (2019)

Bibliography

Adams, D & Carwardine, M (2009) *Last Chance to See*. Arrow.

Aebischer, T, Ibrahim, T, Hickisch, R et al (2020) Apex predators decline after an influx of pastoralists in former Central African Republic hunting zones. *Biological Conservation* 241: 108326.

Antunes, A, Troyer, JL, Roelke, ME et al (2008) The evolutionary dynamics of the lion (*Panthera leo*) revealed by host and viral population genomics. *PLOS Genetics* 4: e1000251.

Barnett, R, Sinding, M-HS, Vieira, FG et al (2018) No longer locally extinct? Tracing the origins of a lion (*Panthera leo*) living in Gabon. *Conservation Genetics* 19: 611–618.

Barnett, R, Yamaguchi, N, Barnes, I et al (2006) The origin, current diversity and future conservation of the modern lion (*Panthera leo*). *Proceedings of the Biological Society* 273: 2119–2125.

Barnett, R, Yamaguchi, N, Shapiro, B et al (2014) Revealing the maternal demographic history of *Panthera leo* using ancient DNA and a spatially explicit genealogical analysis. *BMC Evolutionary Biology* 14: 70.

Barton, K (2020) MuMIn: Multi-Model Inference. R package version 1.43.17. Available at: https://CRAN.R-project.org/package=MuMIn.

Bauer, H, Chapron, G, Nowell, K et al (2015) Lion (*Panthera leo*) populations are declining rapidly across Africa, except in intensively managed areas. *Proceedings of the National Academy of Sciences* 112: 14894–14899.

Bauer, H, Chardonnet, B, Scholte, P et al (2021) Consider divergent regional perspectives to enhance wildlife conservation across Africa. *Nature Ecology and Evolution* 5(2): 149–152.

Bauer, H & De Iongh, H (2005) Lion (*Panthera leo*) home ranges and livestock conflicts in Waza National Park, Cameroon. *African Journal of Ecology* 43(3): 208–214.

Bauer, H, De Iongh, H & Sogbohossou, E (2010) Assessment and mitigation of human-lion conflict in West and Central Africa. *Mammalia* 74: 363–367.

Bauer, H, Packer, C, Funston, PJ et al (2016) *Panthera leo* (errata version published in 2017). The IUCN Red List of Threatened Species 2016: e.T15951A115130419. Available at: https://doi.org/10.2305/IUCN.UK.2016-3.RLTS.T15951A107265605.en.

Bauer, H, Page-Nicholson, S, Hinks, AE et al (2018) Status of the lion in sub-Saharan Africa. In *Guidelines for the Conservation of Lions in Africa*. Unpublished report. IUCN Cat Specialist Group, Muri bei Bern, Switzerland.

Bauer, H & Van der Merwe, S (2004) Inventory of free-ranging lions (*Panthera leo*) in Africa. *Oryx* 38: 26–31.

Becker, MS, Almeida, J, Begg, C et al (2022) Guidelines for evaluating the conservation value of African lion (*Panthera leo*) translocations. *Frontiers in Conservation Science* 3: Available at: https://doi.org/10.3389/fcosc.2022.963961.

Bell, RHV (1982) The effect of soil nutrient availability on the community structure in African ecosystems. In Huntley, BJ & Walker, BH (eds.), *Ecology of Tropical Savannas* (pp. 193–216). Springer-Verlag.

Benson, JF, Mahoney, PJ, Sikich, JA et al (2016) Interactions between demography, genetics, and landscape connectivity increase extinction probability for a small population of large carnivores in a major metropolitan area. *Proceedings of the Royal Society B: Biological Sciences* 283: 20160957.

Bertola, LD, Jongbloed, H, Van der Gaag, KJ et al (2016) Phylogeographic patterns in Africa and high-resolution delineation of genetic clades in the lion (*Panthera leo*). *Scientific Reports* 6: 30807.

Bertola, LD, Miller, SM, Williams, V et al (2021) Genetic guidelines for translocations: Maintaining intraspecific diversity in the lion (*Panthera leo*). *Evolutionary Applications* 15: 22–39.

Bertola, LD, Sogbohossou, EA, Palma, L et al (2022a) Policy implications from genetic guidelines for the translocations of lions. *Cat News* 75: 41–43.

Bertola, LD, Tensen, L, Van Hooft, P et al (2015) Autosomal and mtDNA markers affirm the distinctiveness of lions in West and Central Africa. *PLOS ONE* 11(3): e0149059.

Bertola, LD, Van Hooft, WF, Vrieling, K et al (2011) Genetic diversity, evolutionary history and implications for conservation of the lion (*Panthera leo*) in West and Central Africa. *Journal of Biogeography* 38 (7): 1356–1367.

Bertola, LD, Vermaat, M, Lesilau, F et al (2019) Whole genome sequencing and the application of a SNP panel reveal primary evolutionary lineages and genomic diversity in the lion (*Panthera leo*). bioRxiv: 814103.

Bertola, LD, Vermaat, M, Lesilau, F et al (2022b) Whole genome sequencing and the application of a SNP panel reveal primary evolutionary lineages and genomic variation in the lion (*Panthera leo*). *BMC Genomics* 23: 321.

Bhanusinh, D (2006) Junagadh state and its Lions: Conservation in princely India, 1879–1947. *Conservation and Society* 4(4): 522-540.

Blackwell, BF, DeVault, TL, Fernández-Juricic, E et al (2016) No single solution: Application of behavioural principles in mitigating human-wildlife conflict. *Animal Behaviour* 120: 245–254.

Bongaarts, J (2009) Human population growth and the demographic transition. *Philosophical Transactions of the Royal Society B: Biological Sciences* 364: 2985–2990.

Borgerhoff Mulder, M, Fitzherbert, EB, Mwalyoyo, J et al (2009) *The national Sukuma expansion and Sukuma-lion conflict*. Report prepared for Panthera. Available at: https://watusimbamazingira.wordpress.com/wp-content/uploads/2012/07/panthera-report-2009_sukuma-expansion-lion-killing1.pdf.

Borgerhoff Mulder, M, Kwiyega, JL, Beccaria, S et al (2019) Lions, Bylaws and Conservation Metrics. *BioScience* 69(12): 1008–1018. Available at: https://academic.oup.com/bioscience/article/69/12/1008/5606896.

Born Free Foundation (2008) *Too much pressure to handle? Lion derivatives in traditional medicine in Nigeria, West Africa*. Horsham: Born Free Foundation.

Borrego, N & Gaines, M (2016) Social carnivores outperform asocial carnivores on an innovative problem. *Animal Behaviour* 114: 21–26.

Boshoff, A, Landman, M & Kerley, G (2016) Filling the gaps on the maps: Historical distribution patterns of some larger mammals in part of southern Africa. *Transactions of the Royal Society of South Africa* 71: 23–87.

Bouché, P, Douglas-Hamilton, I, Wittemyer, G et al (2011) Will elephants soon disappear from West African Savannahs? *PLOS ONE* 6: e20619.

Bouché, P, Renaud, P-C, Lejeune, P et al (2010) Has the final countdown to wildlife extinction in Northern Central African Republic begun? *African Journal of Ecology* 48: 994–1003.

Bradshaw, CJA & Di Minin, E (2019) Socio-economic predictors of environmental performance among African nations. *Scientific Reports* 9: 9306.

Brink, H, Smith, RJ, Skinner, K et al (2016) Sustainability and long term-tenure: Lion trophy hunting in Tanzania. *PLOS ONE* 11(9): e0162610.

Burger, J, Rosendahl, W, Loreille, O et al (2004) Molecular phylogeny of the extinct cave lion (*Panthera leo spelaea*). *Molecular Phylogenetics and Evolution* 30: 841–849.

Cardillo, M, Mace, GM, Jones, KE et al (2005) Multiple causes of high extinction risk in large mammal species. *Science* 309: 1239–1241.

Ceballos, G & Ehrlich, PR (2002) Mammal population losses and the extinction crisis. *Science* 296: 904–907.

Ceballos, G, Ehrlich, PR & Dirzo, R (2017) Biological annihilation via the ongoing sixth mass extinction signaled by vertebrate population losses and declines. *Proceedings of the National Academy of Sciences, USA* 114: E6089–E6096.

Chardonnet, P (2002) Conservation of the African lion: Contribution to a status survey. International Foundation for the Conservation of Wildlife, France and Conservation Force, USA.

Coe, MJ, Cumming, DH & Phillipson, J (1976) Biomass and production of large African herbivores in relation to rainfall and primary production. *Oecologia* 22: 341–354.

Compton, BW, McGarigal, K, Cushman, SA et al (2007) A resistant-kernel model of connectivity for amphibians that breed in vernal pools. *Conservation Biology* 21: 788–799.

Craigie, ID, Baillie, JEM, Balmford, A et al (2010) Large mammal population declines in Africa's protected areas. *Biological Conservation* 143: 2221–2228.

Crooks, KR (2002) Relative sensitivities of mammalian carnivores to habitat fragmentation. *Conservation Biology* 16: 488–502.

Crooks, KR, Burdett, CL, Theobald, DM et al (2017) Quantification of habitat fragmentation reveals extinction risk in terrestrial mammals. *Proceedings of the National Academy of Sciences* 114: 7635–7640.

Curry, CJ, Davis, BW, Bertola, LD et al (2020) Spatiotemporal genetic diversity of lions reveals the influence of habitat fragmentation across Africa. *Molecular Biology and Evolution* 38(1): 48–57.

Curry, CJ, White, PA & Derr, JN (2019) Genetic analysis of African lions (*Panthera leo*) in Zambia support movement across anthropogenic and geographical barriers. *PLOS ONE* 14: e0217179.

Cushman, SA, Elliot, N, Macdonald, DW et al (2015) A multi-scale assessment of population connectivity in African lions (*Panthera leo*) in response to landscape change. *Landscape Ecology* 31: 1337–1353. Available at: https://doi.org/10.1007/s10980-015-0292-3.

Cushman, SA, Elliot, NB, Bauer, D et al (2018) Prioritizing core areas, corridors and conflict hotspots for lion conservation in southern Africa. *PLOS ONE* 13: e0196213.

Cushman, SA & Landguth, EL (2012) Multi-taxa population connectivity in the Northern Rocky Mountains. *Ecological Modelling* 231: 101–112.

Cushman, SA, Landguth, EL & Flather, CH (2013) Evaluating population connectivity for species of conservation concern in the American Great Plains. *Biodiversity and Conservation* 22: 2583–2605.

Cushman, SA & Lewis, JS (2010) Movement behavior explains genetic differentiation in American black bears. *Landscape Ecology* 25: 1613–1625.

Cushman, SA, Lewis, JS & Landguth, EL (2014) Why did the bear cross the road? Comparing the performance of multiple resistance surfaces and connectivity modeling methods. *Diversity* 6: 844–854.

Cushman, SA & McGarigal, K (2019) Metrics and models for quantifying ecological resilience at landscape scales. *Frontiers in Ecology and Evolution* 7: 440.

Cushman, SA, McKelvey, MK, Fau-Schwartz, KS et al (2009) Use of empirically derived source-destination models to map regional conservation corridors. *Conservation Biology* 23(2): 368–376.

Cushman, SA, McRae, B, Adriaensen, F et al (2013) Biological corridors and connectivity. *Key Topics in Conservation Biology* 2: 384–404.

Delaney, KS, Riley, SPD & Fisher, RN (2010) A rapid, strong, and convergent genetic response to urban habitat fragmentation in four divergent and widespread vertebrates. *PLOS ONE* 5: e12767.

De Manuel, M, Barnett, R, Sandoval-Velasco, M et al (2020) The evolutionary history of extinct and living lions. *Proceedings of the National Academy of Sciences* 117: 10927–10934.

De Vivo, M & Carmignotto, AP (2004) Holocene vegetation change and the mammal faunas of South America and Africa. *Journal of Biogeography* 31: 943–957.

De Waal, L, Jakins, C, Klarmann, SE et al (2022) The unregulated nature of the commercial captive predator industry in South Africa: Insights gained using the PAIA process. *Nature Conservation* 50: 227–264. Available at: https://natureconservation.pensoft.net/article/85108/.

Díaz, S, Settele, J, Brondízio, ES et al (2019) Pervasive human-driven decline of life on Earth points to the need for transformative change. *Science* 366: eaax3100.

Dickman, AJ (2008) *Key determinants of conflict between people and wildlife, particularly large carnivores, around Ruaha National Park, Tanzania.* Page 369. PhD dissertation, Department of Anthropology. University College London, London. Available at: www.researchgate.net/publication/39065771.

Di Minin, E, Slotow, R, Fink, C et al (2021) A pan-African spatial assessment of human conflicts with lions and elephants. *Nature Communications* 12: 2978.

Diniz, MF, Cushman, SA, Machado, RB et al (2020) Landscape connectivity modeling from the perspective of animal dispersal. *Landscape Ecology* 35: 41–58

Divyabhanusinh (2014) Lions, cheetahs, and others in the Mughal landscape. In Rangarajan, M & Sivaramakrishnan, K (eds), *Shifting Ground: People, Animals, and Mobility in India's Environmental History.* Oxford Academic.

Dolrenry, S, Stenglein, J, Hazzah, L et al (2014) A metapopulation approach to African lion (*Panthera leo*) conservation. *PLOS ONE* 9: e88081..

Dominguez, L & Luoma, C (2020) Decolonising conservation policy: How colonial land and conservation ideologies persist and perpetuate indigenous injustices at the expense of the environment. *Land* 9(3), 65. Available at: https://doi.org/10.3390/land9030065

Dubach, JM, Briggs, MB, White, PA et al (2013) Genetic perspectives on 'Lion Conservation Units' in Eastern and Southern Africa. *Conservation Genetics* 14: 741–755.

Dures, SG, Carbone, C, Loveridge, AJ et al (2019) A century of decline: Loss of genetic diversity in a southern African lion-conservation stronghold. *Diversity and Distributions* 25: 870–879.

East, RD (1984) Rainfall, soil nutrient status and biomass of large African savanna mammals. *African Journal of Ecology* 22: 245–270.

Elliot, N (2014) *The ecology of dispersal in lions* (Panthera leo). PhD thesis, Oxford University.

Elliot, N, Cushman, SA, Macdonald, DW et al (2014) The devil is in the dispersers. Predictions of landscape connectivity change with demography. *Journal of Applied Ecology* 51: 1169–1178.

Estrada-Carmona, N, Hart, AK, DeClerck, FAJ, et al (2014) Integrated landscape management for agriculture, rural livelihoods, and ecosystem conservation: An assessment of experience from Latin America and the Caribbean. *Landscape and Urban Planning* 129: 1–11.

Fitzherbert, E, Caro, T, Johnson, PJ et al (2014) From avengers to hunters: leveraging collective action for the conservation of endangered lions. *Biological Conservation* 174: 84–92.

Frank, L, Hemson, G, Kushnir, H et al (2006) Lions, conflict and conservation in Eastern and Southern Africa. Eastern and Southern African Lion Conservation workshop, Johannesburg, South Africa. Available at: www.researchgate.net./publication/242139181.

Goldman, MJ, De Pinho, JR & Perry, J (2013) Beyond ritual and economics: Maasai lion hunting and conservation politics. *Oryx* 47(4): 490–500. Available at: https://doi.org/10.1017/S0030605312000907.

Goldman, MJ, De Pinho, JR & Perry, J (2010) Maintaining complex relations with large cats: Maasai and lions in Kenya and Tanzania. *Human Dimensions of Wildlife* 15(5): 332–346. Available at: http://doi.org/10.1080/10871209.2010.506671.

Goodrich, J, Lynam, A, Miquelle, D et al (2015) *Panthera tigris*. The IUCN Red List of Threatened Species 2015: e.T15955A50659951. Available at: www.researchgate.net/publication/301296266_Panthera_tigris_The_IUCN_Red_List_of_Threatened_Species_2015.

Grange, S & Duncan, P (2006) Bottom-up and top-down processes in African ungulate communities: Resources and predation acting on the relative abundance of zebra and grazing bovids. *Ecography* 29: 899–907.

Green, J, Jakins, C, Asfaw, E et al (2020) African Lions and Zoonotic Diseases: Implications for commercial lion farms in South Africa. *Animals* 10(9): 1692. Available at: https://doi.org/10.3390/ani10091692.

Guggisberg, CAW (1961) *Simba: the life of the lion.* Bailey Bros and Swinfen.

Guisan, A, Edwards, TC Jr & Hastie, T (2002) Generalized linear and generalized additive models in studies of species distributions: Setting the scene. *Ecological Modelling* 157: 89–100.

Harcourt, AH, Parks, SA & Woodroffe, R (2001) Human density as an influence on species/area relationships: Double jeopardy for small African reserves? *Biodiversity and Conservation* 10: 1011–1026.

Hatton, IA, McCann, KS, Fryxell, JM et al (2015) The predator-prey power law: Biomass scaling across terrestrial and aquatic biomes. *Science* 349: aac6284.

Hazzah, L, Borgerhoff Mulder, M & Frank, L (2009) Lions and warriors: Social factors underlying declining African lion populations and the effect of incentive-based management in Kenya. *Biological Conservation* 142: 2428–2437.

Hazzah, LN (2007) *Living Among Lions (*Panthera leo*): Coexistence or Killing? Community Attitudes towards Conservation Initiatives and the Motivation behind Lion Killing in Kenyan Maasai land.* PhD dissertation. University of Wisconsin-Madison, USA.

Hedwig, D, Kienast, I, Bonnet, M et al (2018) A camera trap assessment of the forest mammal community within the transitional savannah-forest mosaic of the Batéké Plateau National Park, Gabon. *African Journal of Ecology* 56: 777–790.

Heinrich, S, Gomez, L, Green, J et al (2022) The extent and nature of the commercial captive lion industry in the Free State province, South Africa. *Nature Conservation* 50: 203-225

Henschel, P, Bauer, H, Sogbohossou, E et al (2015) *Panthera leo* (West Africa subpopulation). The IUCN Red List of Threatened Species 2015: e.T68933833A54067639. Available at: www.cms.int/sites/default/files/document/cites-cms_aci2_inf.13_lion-iucn-red-list_e.pdf.

Henschel, P, Coad, L, Burton, C et al (2014) The lion in West Africa is critically endangered. *PLOS ONE* 9(1): 1–11.

Henschel, PP, Malanda, G & Hunter, L (2014) The status of savanna carnivores in the Odzala-Kokoua National Park, northern Republic of Congo. *Journal of Mammalogy* 95: 882–892.

Holdren, JP & Ehrlich, PR (1974) Human population and the global environment: Population growth, rising per capita material consumption, and disruptive technologies have made civilization a global ecological force. *American Scientist* 62: 282–292.

Holmern, T, Nyahongo, J & Røskaft, E (2007) Livestock loss caused by predators outside the Serengeti National Park, Tanzania. *Biological Conservation* 135: 518–526.

Hunter, LT, White, P, Henschel P et al (2013) Walking with lions: Why there is no role for captive-origin lions (*Panthera leo*) in species restoration. *Oryx* 47(1): 19–24.

Israel, P (2009). The war of lions: witch-hunts, occult idioms and post-socialism in northern Mozambique. *Journal of Southern African Studies* 35:155-174.

IUCN-SSC (2006a) *Conservation Strategy for the Lion in West and Central Africa.* IUCN Species Survival Commission Cat Specialist Group, Muri bei Bern, Switzerland.

IUCN-SSC (2006b) *Conservation Strategy for the Lion in Eastern and Southern Africa.* IUCN Species Survival Commission Cat Specialist Group, Muri bei Bern, Switzerland.

IUCN-SSC (2006) *Conservation strategy for the lion (Panthera leo) in eastern and southern Africa.* IUCN Species Survival Commission Cat Specialist Group, Muri bei Bern, Switzerland.

IUCN-SSC (2018) *Guidelines for the Conservation of Lions in Africa.* Version 1.0 (147). IUCN Species Survival Commission Cat Specialist Group, Muri bei Bern, Switzerland..

Kaszta, Ż, Cushman, SA, Htun, S et al (2020a) Simulating the impact of the Belt and Road initiative and other major developments in Myanmar on an ambassador felid, the clouded leopard *Neofelis nebulosa*. *Landscape Ecology* 35: 727–747.

Kaszta, Ż, Cushman, SA & Macdonald, DW (2020b) Prioritizing habitat core areas and corridors for a large carnivore across its range. *Animal Conservation* 23: 607–616.

Kideghesho, JR (2008) Who pays for wildlife conservation in Tanzania and who benefits? Working paper – 12th Biennial Conference of the International Association for the Study of Commons (Cheltenham). Available at: https://dlc.dlib.indiana.edu/dlc/bitstream/handle/10535/587/Kideghesho_102301.pdf?sequence=1.

Kiffner, C, Meyer, B, Mühlenberg, M et al (2009) Plenty of prey, few predators: What limits lions (*Panthera leo*) in Katavi National Park, western Tanzania? *Oryx* 43: 52–59.

Kingdon, J (2018) *The Kingdon Field Guide to African Mammals: Second Edition.* Bloomsbury Publishing.

Kissui, BM (2008) Livestock predation by lions, leopards, spotted hyaenas, and their vulnerability to retaliatory killing in the Maasai Steppe, Tanzania. *Animal Conservation* 11: 422–432. Available at: http://dx.doi.org/10.1111/j.1469-1795.2008.00199.x.

Kitchener, A, Breitenmoser, C, Eizirik, E et al (2017) A revised taxonomy of the Felidae: The final report of the Cat Classification Task Force of the IUCN Cat Specialist Group. *Cat News* 11 (winter).

Kushnir, H, Leitner, H, Ikanda, D et al (2010) Human and ecological risk factors for unprovoked lion attacks on humans in southeastern Tanzania. *Human Dimensions of Wildlife*. 15. 315–331.

Kwiyega, JL, Mwaja, NS & Gilya, L (2020) WASIMA Campaign: Wildlife Law Enforcement Stakeholders Workshops Report for Stopping Illegal Lion Killing in Western Tanzania. Landscape and Conservation Mentors Organization (LCMO). Available at: https://ruffordorg.s3.amazonaws.com/media/project_reports/29894-2-Final_Evaluation_Report.pdf.

Landguth, EL, Hand, BK, Glassy, J et al (2012) UNICOR: A species connectivity and corridor network simulator. *Ecography* 35: 9–14.

Laurance, WF, Clements, GR, Sloan, S et al (2014) A global strategy for road building. *Nature* 513: 229–232.

Lhoest, S, Linchant, J, Gore, ML et al (2022) Conservation science and policy should care about violent extremism. *Global Environmental Change* 76: 102590.

Liesegang, G (2003) Survey of the history of the population falling under the districts of Mavago and Mecula and the areas enclosed within the Niassa Game Reserve. Unpublished report for SRN, Maputo.

Lindsey, PA, Balme, GA, Funston, P et al (2013) The trophy hunting of African lions: Scale, current management practices and factors undermining sustainability. *PLOS ONE* 8(9): e73808.

Lindsey, PA, Miller, JRB, Petracca, LS et al (2018) More than $1 billion needed annually to secure Africa's protected areas with lions. *Proceedings of the National Academy of Sciences* 115: e10788–e10796.

Lindsey, PA, Petracca, LS, Funston, PJ et al (2017) The performance of African-protected areas for lions and their prey. *Biological Conservation* 209: 137–149.

Loveridge, AJ & Canney, S (2009) Report on the lion distribution and conservation modelling project. Unpublished Report to the Born Free Foundation, UK.

Loveridge, AJ, Sousa, LL, Cushman, S et al (2022) Where have all the lions gone? Establishing realistic baselines to assess decline and recovery of African lions. *Diversity and Distributions* 28: 2388–2402. Available at: https://doi.org/10.1111/ddi.13637.

Loveridge, AJ, Valeix, M, Elliot, NB et al (2017) The landscape of anthropogenic mortality: How African lions respond to spatial variation in risk. *Journal of Applied Ecology* 54: 815–825.

Loveridge, AJ, Wang, SW, Frank, L et al (2010) People and wild felids: Conservation of cats and management of conflicts. In Macdonald, DW and Loveridge, AJ (eds), *Biology and Conservation of Wild Felids*: 161–195. Oxford University Press.

Macdonald, DW, Chiaverini, L, Bothwell, HM et al (2020) Predicting biodiversity richness in rapidly changing landscapes: Climate, low human pressure or protection as salvation? *Biodiversity and Conservation* 29: 4035–4057.

Macdonald, E, Burnham, D, Hinks, AE et al (2015) Conservation inequality and the charismatic cat: *Felis felicis. Global Ecology and Conservation* 3: 851–866.

Maclennan, SD, Groom, JR, Macdonald, DW et al (2009) Evaluation of a compensation scheme to bring about pastoralist tolerance of lions. *Biological Conservation* 142(11): 2419–2427.

Main, D (2020) Why we don't really know how many lions live in Africa. *National Geographic*. Available at: www.nationalgeographic.com/animals/article/how-african-lion-populations-are-estimated.

Marchini, S, Ferraz, K, Zimmermann, A et al (2019) Planning for coexistence in a complex human-dominated world. In Frank, B, Glikman, J & Marchini, S (eds), *Human–wildlife interactions: Turning conflict into coexistence*: 414–438. Cambridge University Press.

Matshisela, A, Elliot, N, Chinoitezvi, E et al (2021) Long distance African lion dispersal between two protected areas. *African Journal of Ecology* 60: 67–70.

McGarigal, K, Cushman, SA & Ene, E (2012) FRAGSTATS v4: Spatial pattern analysis program for categorical and continuous maps. Computer software programme produced by the authors at the University of Massachusetts.

Mésochina, P, Mbangwa, O, Chardonnet, P et al (2010) *Conservation status of the lion (Panthera leo Linnaeus, 1758) in Tanzania*. SCI Foundation, MNRT-WD, TAWISA and IGF Foundation, Paris, France. 113pp.

Miller, SM & Funston, PJ (2014) Rapid growth rates of lion (*Panthera leo*) populations in small, fenced reserves in South Africa: A management dilemma. *South African Journal of Wildlife Research* 44: 43–55.

Miller, SM, Harper, CK, Bloomer, P et al (2015) Fenced and fragmented: Conservation value of managed metapopulations. *PLOS ONE* 10(12): e0144605.

Milton, K (1997) Ecologies: Anthropology culture, and the environment. *International Social Science Journal*. 49: 477–495

Mohammed, AA, Bauer, H, Coals, P et al (2023) Abundance of larger mammals in Dinder National Park, Sudan. *European Journal of Wildlife Research* 69(3): 46.

Mole, KH & Newton, D (2020) *An assessment of trade, mortalities and anthropogenic threats facing lions in Tanzania and Mozambique*. TRAFFIC International, Cambridge, United Kingdom. Available at: www.traffic.org/publications/reports/african-lion-trade-an-assessment-of-trade-mortalities-and-anthropogenic-threats-facing-lions-in-tanzania-and-mozambique/.

Morandin, C, Loveridge, AJ, Segelbacher, G et al (2014) Gene flow and immigration: Genetic diversity and population structure of lions (*Panthera leo*) in Hwange National Park, Zimbabwe. *Conservation Genetics* 15: 697–706.

Morapedi, M, Reuben, M, Gadimang, P et al (2021) Outcomes of lion (*Panthera leo*) translocations to reduce conflict with farmers in Botswana. *African Journal of Wildlife Research* 51: 6–12.

Morrison, JC, Sechrest, W, Dinerstein, E et al (2007) Persistence of large mammal faunas as indicators of global human impacts. *Journal of Mammalogy* 88: 1363–1380.

Myers, N (1975) The silent savannahs. *International Wildlife* 5: 5–10.

Newitt, M (1995) *A History of Mozambique*. Indiana Press University.

Newmark, WD (2008) Isolation of African protected areas. *Frontiers in Ecology and the Environment* 6: 321–328.

Nowell, K & Jackson, P (1996) *Wild cats: A status survey and conservation action plan*. IUCN Species Survival Commission Cat Specialist Group, Muri bei Bern, Switzerland.

O'Connor, VL, Thomas, P, Chodorow, M et al (2022) Exploring innovative problem-solving in African lions (*Panthera leo*) and snow leopards (*Panthera uncia*). *Behavioural Processes* 199: 1046–1048.

Ogada, M, Woodroffe, R, Oguge, N et al (2003) Limiting depredation by African carnivores. *Conservation Biology* 17: 1521–1530.

Oksanen, J, Blanchet, FG, Friendly, M et al (2020) Vegan: Community ecology package. R package version 2.5-7. Available at: https://CRAN.R-project.org/package=vegan.

Olléová, M & Dogringar, S (2013) *Carnivore Monitoring Program Zakouma National Park*. Zakouma National Park, Chad. Unpublished report.

Onorato, DP, Belden, C, Cunningham, MW et al (2010) Long-term research on the Florida panther (*Puma concolor coryi*): Historical findings and future obstacles to population persistence. In Macdonald, DW & Loveridge, AJ (eds.), *Biology and Conservation of Wild Felids*: 454–469. Oxford University Press.

Östberg, W, Howland, O, Mduma, J et al (2018) Tracing improving livelihoods in rural Africa using local measures of wealth: A case study from Central Tanzania, 1991–2016. *Land* 7(2): 44.

Packer, C, Brink, H, Kissui, BM et al (2011) Effects of trophy hunting on lion and leopard populations in Tanzania. *Conservation Biology* 25: 142–153.

Packer, C, Ikanda, D, Kissui, B et al (2006) The ecology of man-eating lions in Tanzania. In Laverdiere, M (ed.), *Human–Wildlife Conflicts*: 10–14. Food and Agriculture Organization of the United Nations, Accra, Ghana.

Packer, C, Ikanda, D, Kissui, B et al (2005) Lion attacks on humans in Tanzania. *Nature* 436: 927–928.

Packer, C, Pusey, A, Eberly, LE (2001) Egalitarianism in female African lions. *Science* 293: 690–693.

Packer, C, Pusey, AE, Rowley, H et al (1991) Case study of a population bottleneck: Lions of the Ngorongoro Crater. *Conservation Biology* 5: 219–230.

Panthera (2021) Meet the lion. Available at: www.panthera.org/cat/lion.

Patterson, BD, Kasiki, SM, Selmpo, E et al (2004) Livestock predation by lions (*Panthera leo*) and other carnivores on ranches neighboring Tsavo National parks, Kenya. *Biological Conservation* 119: 507–516. Available at: www.sciencedirect.com/science/article/abs/pii/S0006320704000163.

Pauly, D (1995) Anecdotes and the shifting baseline syndrome of fisheries. *Trends in Ecology & Evolution* 10: 430.

Pinnock, D & Bell, C (2019) *The Last Elephants*. Struik Nature.

Plumeridge, AA & Roberts, CM (2017) Conservation targets in marine protected area management suffer from shifting baseline syndrome: A case study on the Dogger Bank. *Marine Pollution Bulletin* 116: 395–404.

Ram, M, Sahu, A, Srivastava, N et al (2023) Diet composition of Asiatic lions in protected areas and multi-use land matrix. *Journal of Vertebrate Biology* 72(22065): 1–9.

Reed, DH & Frankham, R (2003) Correlation between fitness and genetic diversity. *Conservation Biology* 17: 230–237.

Riggio, J, Jacobson, A, Dollar, L et al (2013) The size of savannah Africa: A lion's (*Panthera leo*) view. *Biodiversity and Conservation* 22: 17–35.

Ripple, WJ, Estes, JA, Beschta, RL et al (2014) Status and ecological effects of the world's largest carnivores. *Science* 343: 124–148.

Robson, A, Trimble, M, Bauer, DT et al (2021) Over 80% of Africa's savannah conservation land is failing or deteriorating according to lions as an indicator species. *Conservation Letters* 15: e12844.

Roman, J, Dunphy-Daly, MM, Johnston, DW et al (2015) Lifting baselines to address the consequences of conservation success. *Trends in Ecology & Evolution* 30: 299–302.

Rookmaaker, K & Antoine, P (2012) New maps representing the historical and recent distribution of the African species of rhinoceros: *Diceros bicornis*, *Ceratotherium simum* and *Ceratotherium cottoni*. *Pachyderm* 52: 91–96.

Schapira, P, Monico, M, Rolkier, G, & Bauer, H (2016) Wildlife migration in Ethiopia and South Sudan longer than 'the longest in Africa': a response to Naidoo et al. *Oryx*. 51: 1.

Schnitzler, AE (2011) Past and present distribution of the North African-Asian lion subgroup: A review. *Mammal Review* 41: 220–243.

Schroth, G & McNeely, JA (2011) Biodiversity conservation, ecosystem services and livelihoods in tropical landscapes: Towards a common agenda. *Environmental Management* 48: 229–236.

Sibanda, L, Johnson, P, Van der Meer, E et al (2021) Effectiveness of community-based livestock protection strategies: A case study of human-lion conflict mitigation. *Oryx* 56: 537–545.

Smitz, N, Jouvenet, O, Ambwene Ligate, F et al (2018) A genome-wide data assessment of the African lion (*Panthera leo*) population genetic structure and diversity in Tanzania. *PLOS ONE* 13: e0205395.

Somerville, K (2020) *Human and Lions: Conflict, Conservation and Coexistence*. Routledge.

Spear, T & Waller, R (1993) *Being Maasai: Ethnicity & Identity in East Africa*. James Currey Publishers.

Stuart, AJ & Lister, AM (2011) Extinction chronology of the cave lion Panthera spelaea. *Quaternary Science Review* 30: 2329–2340.

Tanzania Wildlife Research Institute (2009) Tanzania Lion and Leopard Conservation Action Plan. In *Tanzania Carnivore Conservation Action Plan*: 64–111. TAWIRI, Arusha, Tanzania.

Tende, T, Hansson, B, Ottosson, U et al (2014) Individual identification and genetic variation of lions (*Panthera leo*) from two protected areas in Nigeria. *PLOS ONE* 9: e84288.

Tilman, D, Clark, M, Williams, DR et al (2017) Future threats to biodiversity and pathways to their prevention. *Nature* 546: 73–81.

Tree, I (2019) *Wilding: The Return of Nature to a British Farm*. Picador.

Trinkel, M, Cooper, D, Packer, C et al (2011) Inbreeding depression increases susceptibility to Bovine tuberculosis in lions: An experimental test using an inbred-outbred contrast through translocation. *Journal of Wildlife Diseases* 47(3): 494–500.

Trinkel, M, Ferguson, N, Reid, A et al (2008) Translocating lions into an inbred lion population in the Hluhluwe-iMfolozi Park, South Africa. *Animal Conservation* 11: 138–143.

United Nations (2019) *World population prospects 2019, Volume II: Demographic profiles* (ST/ESA/SER.A/427). United Nations. Department of Economic and Social Affairs.

Van de Kerk, M, Onorato, DP, Hostetler, JA et al (2019) Dynamics, Persistence, and Genetic Management of the Endangered Florida Panther Population. *Wildlife Monographs* 203(1): 3–35.

Vanherle, N (2011) Inventaire et suivi de la population de lions (*Panthera leo*) du Parc National de Zakouma (Tchad). *Revue d'Écologie (La Terre et la Vie)* 66: 317–366.

Van Orsdol, KG, Hanby, JP & Bygott, JD (1985) Ecological correlates of lion social organization (*Panthera leo*). *Journal of Zoology* 206: 97–112.

Vettorazzi, M, Mogensen, N, Kaelo, B & Broekhuis, F (2022) Understanding the effects of seasonal variation in prey availability on prey switching by large carnivores. *Journal of Zoology* 318(3): 218–227.

Wasser, S, Poole, J, Lee, P et al (2010) Conservation. Elephants, ivory, and trade. *Science* 327: 1331–1332.

Whitman, KL, Starfield, AM, Quadling, H et al (2007) Modeling the effects of trophy selection and environmental disturbance on a simulated population of African Lions. *Conservation Biology* 21: 591–601. Available at: https://doi.org/10.1111/j.1523-1739.2007.00700.x.

Wildt, DE, Bush, M, Goodrowe, KL et al (1987) Reproductive and genetic consequences of founding isolated lion populations. *Nature* 329: 328–331.

Williams, VL, Loveridge, AJ, Newton, DJ et al (2017) Questionnaire survey of the pan-African trade in lion body parts. *PLOS ONE* 12(10): e0187060.

Wilson, GM (1953) The Tatoga of Tanganyika, part 2. *Tanganyika Notes and Records* 34: 35–56.

Wittemyer, G, Elsen, P, Bean, WT et al (2008) Accelerated human population growth at protected area edges. *Science* 321: 123–126.

Wood, SN (2011) Fast stable restricted maximum likelihood and marginal likelihood estimation of semiparametric generalized linear models. *Journal of the Royal Statistical Society* (B) 73: 3–36.

Woodroffe, R & Ginsberg, JR (1998) Edge effects and the extinction of populations inside protected areas. *Science* 280: 2126–2128.

Yamaguchi, N, Cooper, A, Werdelin, L et al (2004) Evolution of the mane and group-living in the lion (*Panthera leo*): A review. *Journal of Zoology* 263: 329–342.

Zuur, AF, Ieno, EN, Walker, N et al (2009) *Mixed effects models and extensions in ecology with R*. Springer Science & Business Media.

© 2025 by Struik Nature (a division of Penguin Random House South Africa (Pty) Ltd).

All rights reserved. No part of this publication may be reproduced or transmitted in any form or by any means, electronic or mechanical, including photocopying, recording, or information storage or retrieval system, without permission in writing from the publishers.

Published in North America by Smithsonian Books
PO Box 37012, MRC 513
Washington, DC 20013
smithsonianbooks.com

Director: Carolyn Gleason
Senior Editor: Jaime Schwender
Production Editor: Julie Huggins

Edited by Helen de Villiers
Designed by Verena Altern and Neil Bester

This book may be purchased for educational, business, or sales promotional use. For information, please write the Special Markets Department at the address or website above.

Library of Congress Cataloging-in-Publication Data available upon request.

Paperback ISBN: 978-1-58834-805-0

Printed in China, not at government expense

29 28 27 26 25   1 2 3 4 5

For permission to reproduce illustrations appearing in this book, please correspond directly with the owners of the works, as seen in photography and image credits. Smithsonian Books does not retain reproduction rights for these images individually or maintain a file of addresses for sources.